Principles and Practices of Transportation Planning and Engineering

Principles and Practices of Transportation Planning and Engineering

A Foundational Reference for Professionals and Textbook for Students Focusing on Civil & Transportation Engineering, Planning, Environmental Science, and Project Management

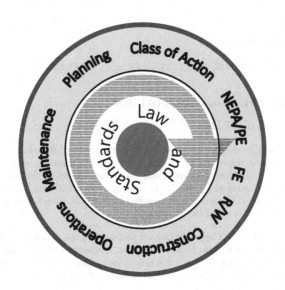

Connie Kelly Tang and Lei Zhang

CRC Press

Taylor & Francis Group
Boca Raton London New York

CRC Press is an imprint of the
Taylor & Francis Group, an **informa** business

First edition published 2021
by CRC Press
6000 Broken Sound Parkway NW, Suite 300, Boca Raton, FL 33487-2742

and by CRC Press
2 Park Square, Milton Park, Abingdon, Oxon, OX14 4RN

ISBN: 978-0-367-70238-0 (hbk)
ISBN: 978-0-367-71474-1 (pbk)
ISBN: 978-1-003-14517-2 (ebk)

Typeset in Garamond
by codeMantra

Contents

Foreword...xvii
Preface xix
Authors xxi

1 The Law and the Process...1
    1.1   The Law...2
          1.1.1   Intermodal Surface Transportation Efficiency Act
                   (ISTEA) 2
          1.1.2   Transportation Equity Act for the 21st Century (TEA-21) ...3
          1.1.3   Safe, Accountable, Flexible, Efficient Transportation
                   Equity Act: A Legacy for Users (SAFETEA-LU) 4
          1.1.4   Moving Ahead for Progress in the 21st Century Act
                   (MAP21)...11
          1.1.5   Fixing America's Surface Transportation (FAST) Act.........12
          1.1.6   Law, Programs, and Projects 17
    1.2   Other Programs...17
          1.2.1   Continuing Resolution 17
          1.2.2   The University Transportation Centers Program..............17
          1.2.3   Business Opportunities with the U.S. Government 18
          1.2.4   Hierarchy of Laws and Regulations.....................................18
          1.2.5   Federal Highway Standards 18
    1.3   Transportation Project Development Process.....................................19
          1.3.1   Basics of the Development Process.....................................19
          1.3.2   Government Agencies and Organizations Involved............21
    1.4   Professional Disciplines 23
          1.4.1   Civil Engineering Licensing...23
          1.4.2   License Reciprocity or Comity...24
          1.4.3   Planning Professionals 24
          1.4.4   Environmental Science...24
    1.5   Summary...25
    1.6   Discussion 25
    1.7   Exercises 25
    Bibliography 26

2 Transportation Planning..**29**
  2.1 Basic Concepts 29
      2.1.1 Roadway Classification 29
            2.1.1.1 Ownership-Based Approach..............................29
            2.1.1.2 Functionality-Based Approach 30
            2.1.1.3 Administrative Approach 35
      2.1.2 Geographical and Statistical Reporting Area Designation....37
            2.1.2.1 Urban Area 37
            2.1.2.2 Rural Areas 37
            2.1.2.3 Metropolitan and Micropolitan Statistical
                    Areas...37
            2.1.2.4 Metropolitan Planning Area 38
            2.1.2.5 Transportation Management Area 38
      2.1.3 National Air Quality Area Designation 38
            2.1.3.1 Non-Attainment Area 38
            2.1.3.2 Maintenance Area 40
  2.2 Transportation Planning Organizations 40
      2.2.1 Metropolitan Planning Organization.................................. 40
            2.2.1.1 Legal Basis 40
            2.2.1.2 MPO Board 40
            2.2.1.3 Voting Right 41
            2.2.1.4 Advisory Committees..41
            2.2.1.5 Professional Staff.. 42
      2.2.2 State Department of Transportation 42
      2.2.3 The U.S. Department of Transportation 43
  2.3 Transportation Planning Product 44
      2.3.1 Metropolitan Planning Organization Deliverable 44
      2.3.2 State DOT Deliverable 45
      2.3.3 U.S. DOT Deliverable 45
      2.3.4 TIP Illustration 45
  2.4 Transportation Planning Procedure... 46
      2.4.1 Planning Principle..48
            2.4.1.1 Public and Community Involvement 48
            2.4.1.2 Relevant Governmental Agencies,
                    Organizations, and Businesses Involvement 48
            2.4.1.3 Professional Leadership 48
      2.4.2 Planning Steps 48
            2.4.2.1 Inventory of Existing Facilities and Services........48
            2.4.2.2 Monitoring both Infrastructure Operating
                    Condition and Performance 49
            2.4.2.3 Involving Public 49
            2.4.2.4 Gathering Land Use Data and Working with
                    Local Zoning Officials 49

2.4.2.5 Gathering Social, Demographic, Economic,
and Other Related Data 49
2.4.2.6 Analyzing Transportation Demand through
Modeling 51
2.4.2.7 Regional Air Quality Issue Consideration 51
2.4.2.8 Cost and Revenue Estimation 51
2.4.2.9 Assigning Appropriate Fund to a Project..............52
2.4.2.10 LRTP and TIP Adoption 52
2.4.3 Planning Regulation 52
2.5 Traffic Safety Data Analysis..53
2.5.1 Crash Data...53
2.5.2 Roadway Characteristic Data...54
2.5.3 Roadway Traffic Data 55
2.5.4 Safety Data Analysis 55
2.5.4.1 Total Crash Analysis 55
2.5.4.2 Normalized Crash Data Analysis 57
2.5.4.3 Geospatial Analysis and Visualization...................57
2.5.4.4 Integrated Safety Data Analysis............................61
2.5.5 Planning Crash Data Analysis Example 61
2.5.6 Highway Safety Improvement Program 64
2.5.7 Traffic Records Coordinating Committee............................. 64
2.6 Travel Demand Analysis.. 64
2.6.1 Concepts and Definitions 65
2.6.1.1 Transportation Corridor 65
2.6.1.2 Roadway Network: Link and Node......................65
2.6.1.3 Roadway Link and Node Capacity........................67
2.6.1.4 Traffic Flow Measurement and Characterization...67
2.6.1.5 Travel Origin and Destination 68
2.6.1.6 Traffic Analysis Zones (TAZs) 68
2.6.1.7 Centroid and Centroid Connector 68
2.6.2 Classic Travel Demand Modeling 70
2.6.3 Trip Generation, Distribution, and Routing 82
2.6.4 Activity-Based Travel Demand Modeling 84
2.7 Transportation Conformity Analysis 84
2.7.1 State Implementation Plan 85
2.7.2 Emission Inventory 88
2.7.3 Transportation Emission Estimation....................................88
2.7.4 Conformity Determination...92
2.7.5 Consequences of Not Meeting Conformity Determination ...92
2.8 Financial Planning 93
2.8.1 Project Cost Estimate...93
2.8.2 Revenue Estimates 93
2.8.3 Project Funding and Sequencing..97

2.9 Next Step...97
2.10 Summary..97
2.11 Discussion 98
2.12 Exercises 98
Bibliography 101

3 Project Environmental Class of Action (COA) Determination 103
3.1 National Environmental Policy Act...104
3.2 Parties Involved in the NEPA Process...105
3.3 Class of Action Determination 106
3.4 Environmental Impact Statement..106
3.5 Categorical Exclusions 107
3.6 Environmental Assessment 108
3.7 Engineering Alternatives for Environmental Analysis...................109
3.8 Next Step..109
3.9 Summary...109
3.10 Discussion 110
3.11 Exercises 110
Bibliography 111

4 Preliminary Engineering...113
4.1 Fundamental Concepts...114
    4.1.1 Basic Roadway Components and Terminologies 114
        4.1.1.1 Traveled Way and Travel Lane 114
        4.1.1.2 Median 114
        4.1.1.3 Shoulder..114
        4.1.1.4 Rumble Strip...115
        4.1.1.5 Curb 115
        4.1.1.6 Curb and Gutter 116
        4.1.1.7 Roadway Ditch 116
        4.1.1.8 Guardrail 116
        4.1.1.9 Bicycle Lane 117
        4.1.1.10 Sidewalk...117
        4.1.1.11 Lane Delineation Mark and Marker..............117
        4.1.1.12 Overhead Roadway Signs...............................117
        4.1.1.13 Roadside Signs 118
        4.1.1.14 Variable Message Sign 118
        4.1.1.15 Driveway...119
        4.1.1.16 Median Opening...119
        4.1.1.17 Utilities 119
        4.1.1.18 Crosswalk 119
        4.1.1.19 Noise Abatement Wall 120
        4.1.1.20 Landscaping Greeneries 120

                   4.1.1.21 Stormwater Retention and Detention Ponds 120

     4.1.2   Station – Distance Measuring...121

     4.1.3   Tangents, Horizontal Curves, and Vertical Curves 122

                   4.1.3.1   Tangents 122

                   4.1.3.2   Horizontal Curves....................................122

                   4.1.3.3   Vertical Curves...127

     4.1.4   Sight Distance..129

     4.1.5   Slope/Grade Effects on Stopping Sight Distance 131

                   4.1.5.1   Scenario 1: Uphill Travel.........................131

                   4.1.5.2   Scenario 2: Downhill Travel 132

4.2   Appropriate Horizontal and Vertical Curve Design......................133

     4.2.1   Design Appropriate Horizontal Curve 133

     4.2.2   Superelevation 134

     4.2.3   Design Appropriate Vertical Curve 137

                   4.2.3.1   Crest Vertical Curve.................................137

                   4.2.3.2   Sag Vertical Curve 138

     4.2.4   Practical Design Consideration..139

                   4.2.4.1   Superelevation...139

                   4.2.4.2   Curve Radius 139

                   4.2.4.3   Tire and Pavement Transverse Side Friction

                               Coefficient..140

                   4.2.4.4   Superelevation Transition Rate.................140

                   4.2.4.5   Vertical Curve..140

     4.2.5   Measuring Sight Distance on a Vertical Curve..................141

4.3   Roadway and Bridge Cross Section Design 141

     4.3.1   Cross Section – Travel Lanes and Roadside Elements 141

     4.3.2   Bridge Basics...145

     4.3.3   Typical Sections 148

                   4.3.3.1   Urban Typical Section.............................148

                   4.3.3.2   Rural Typical Section 150

4.4   Design Control..150

     4.4.1   Design Vehicle 150

     4.4.2   Roadway Functional Class 150

     4.4.3   Traffic 151

                   4.4.3.1   Understanding AADT and Its Relationship

                                 with Design Traffic 153

                   4.4.3.2 K_{30} – Design Hour Factor 155

                   4.4.3.3 D_{30} – D Factor 155

                   4.4.3.4   Number of Lanes Need Estimation...........155

                   4.4.3.5   Percentages of Medium Truck and

                                 Combination Truck 156

                   4.4.3.6   Generalized Level of Service Lookup Table........157

     4.4.4   Design Speed 158

        4.4.5   Access Management 158
        4.4.6   Pedestrian and Bicyclist 158
    4.5   Preliminary Engineering Analysis and Design 158
        4.5.1   Horizontal Alignment 159
        4.5.2   Vertical Alignment 159
        4.5.3   Typical Sections 161
        4.5.4   Design Alternatives 163
        4.5.5   Environmental Impact Evaluation 163
    4.6   Engineering Product and Deliverable 163
    4.7   Next Step 165
    4.8   Summary 165
    4.9   Discussions 166
    4.10 Exercises 166
    Bibliography 170

5   Environmental Analysis 171
    5.1   Human, Social, and Cultural Impact Evaluation 172
        5.1.1   Land-Use Changes 172
        5.1.2   Community Cohesion 173
        5.1.3   Aesthetic 174
        5.1.4   Relocation 174
        5.1.5   Community Services 175
        5.1.6   Historic and Archaeological Resources 175
        5.1.7   Parks and Recreation Areas 176
    5.2   Natural Environmental Impact Evaluation 177
        5.2.1   Wetland 177
        5.2.2   Wildlife and Habitat 179
              5.2.2.1   Action Area 179
              5.2.2.2   Incidental Take 180
        5.2.3   Floodplains, Stormwater Runoff, and Water Quality 180
              5.2.3.1   Stormwater Runoff Quantity (Flooding)
                       Control 180
              5.2.3.2   Stormwater Water Quality Control 181
        5.2.4   Farmland 183
    5.3   Physical Environmental Impact Evaluation 183
        5.3.1   Highway Noise 183
              5.3.1.1   Basic Noise Science 186
              5.3.1.2   Noise Impact Analysis Practice 190
        5.3.2   Air Quality 191
              5.3.2.1   Basic Terminology and Concepts 192
              5.3.2.2   Dispersion Mechanism 192
              5.3.2.3   Gaussian Dispersion Modeling 194

           5.3.2.4   Practices in Transportation Air Quality
                      Project-Level Modeling 194
           5.3.2.5   Greenhouse Gas Emission Estimation............... 195
       5.3.3   Contamination Assessment... 195
           5.3.3.1   Contamination Assessment Objective 196
           5.3.3.2   Contamination Assessment Processes................. 196
           5.3.3.3   Basic Remediation... 197
       5.3.4   Utility Assessment.. 197
           5.3.4.1   Utility Assessment Objective............................ 197
           5.3.4.2   Assessing Procedure 198
           5.3.4.3   Utility Impact Assessment Process 198
  5.4   Final Products and Deliverable.. 199
  5.5   Next Step..201
  5.6   Summary..201
  5.7   Discussion 204
  5.8   Exercises 204
  Bibliography 206

6   Final Engineering Design...207
  6.1   Final Engineering Design Deliverables... 208
       6.1.1   Roadway Plan 208
       6.1.2   Structure Plan...212
       6.1.3   Sign and Pavement Marking Plan 214
       6.1.4   Street Lighting Plan 214
       6.1.5   Intelligent Transportation System (ITS) Plan....................214
       6.1.6   Traffic Signalization and Control Plan 217
       6.1.7   Landscaping Plan...217
       6.1.8   Proposed Right of Way Plan 217
       6.1.9   Environmental Permits and Other Permits 217
       6.1.10 Standard Pay Item...219
       6.1.11 Commitments... 220
  6.2   Design Standard and Specification 220
       6.2.1   General Design Specification 220
       6.2.2   Bridge Design Specification 221
       6.2.3   Pavement Design Specification...224
           6.2.3.1   Asphalt Concrete Pavement.............................. 226
           6.2.3.2   Rigid Pavement 226
       6.2.4   Pavement Marking... 228
       6.2.5   Roadway Signs 231
  6.3   Design Exception 231
  6.4   Constructability Review 233
  6.5   Land Surveying 234

	6.5.1	Control Surveys	234
	6.5.2	Right of Way Surveying	235
	6.5.3	Engineering Design Surveying	235
	6.5.4	Construction Surveying	236
	6.5.5	Staking a Horizontal Curve	237
	6.5.6	Labeling Stakes	239
6.6		Next Step	241
6.6		Summary	241
6.8		Discussion	241
6.9		Exercises	242
		Bibliography	244

7 Right of Way **245**

7.1		Right of Way Definition	246
7.2		Acquiring Right of Way	246
7.3		Acquisition Procedure	246
	7.3.1	Involving Affected Property Owners	247
	7.3.2	Official Notification	247
	7.3.3	Official Appraisal	247
	7.3.4	Written Offer	247
	7.3.5	Purchasing	248
	7.3.6	Mediation or Condemnation	248
7.4		Relocation	248
7.5		Final Deliverable	249
7.6		Next Step	249
7.7		Summary	249
7.8		Discussion	249
7.9		Exercises	250
		Bibliography	251

8 Construction **253**

8.1		Construction Personnel	254
	8.1.1	Land Surveyors	254
	8.1.2	Equipment for Both On- and Off-Road Operations	255
		8.1.2.1 Bulldozer	255
		8.1.2.2 Excavator	255
		8.1.2.3 Loader	256
		8.1.2.4 Motor Grader	257
		8.1.2.5 Compactor/Roller	257
		8.1.2.6 Pavers	258
		8.1.2.7 Portland Cement Concrete Transport Truck	258
	8.1.3	Skilled Labor	259
	8.1.4	Flagger	259
	8.1.5	Construction Engineer	260

8.2 Construction Contract Administration 261
    8.2.1 Pre-bid: Bid Package Preparation 262
    8.2.2 Contractor Prequalification..263
    8.2.3 Contractor Bonding Requirement....................................263
    8.2.4 Contract Letting..264
    8.2.5 Bid Advertisement...264
    8.2.6 Bid Opening 264
    8.2.7 Bid Award 265
8.3 Construction Engineering and Inspection............................265
    8.3.1 Post-Award Management 265
    8.3.2 Change of Contracts..266
8.4 Resolving Construction Dispute..267
    8.4.1 Negotiation...267
    8.4.2 Mediation 268
    8.4.3 Dispute Review Board 268
    8.4.4 Minitrials...268
    8.4.5 Arbitration 268
    8.4.6 Private Judging...268
8.5 Highway Construction Safety 268
    8.5.1 Construction Safety and Standard 269
    8.5.2 Work Zone Safety...269
    8.5.3 Worker High-Visibility Safety Apparel............................271
8.6 Next Step...272
8.7 Summary..273
8.8 Discussion 273
8.9 Exercises 273
Bibliography 276

9 Operations...277
9.1 Signal and Optimization 278
    9.1.1 Traffic Signal Control System 278
    9.1.2 Types of Traffic Signal Control 278
        9.1.2.1 Pre-timed Isolated Controller..............................278
        9.1.2.2 Pre-timed Coordinated Controller 278
        9.1.2.3 Semi-Actuated Traffic Controller 279
        9.1.2.4 Fully Actuated Traffic Controller 279
        9.1.2.5 Interconnected Coordinated Traffic Controller....279
    9.1.3 Traffic Control Phase Plan 279
        9.1.3.1 Vehicle Movement...280
        9.1.3.2 Signal Phasing...281
    9.1.4 Labeling of Intersection Traffic Movement 283
    9.1.5 Traffic Signal Timing Basics 284
        9.1.5.1 Effective Green...284

9.1.5.2   Critical Lane Group 284
9.1.5.3   Dilemma Zone 284
9.1.5.4   Adequate Yellow Consideration 285
9.1.5.5   Minimum Green Consideration 285
9.1.5.6   Maximum Green Consideration 287
9.2   Roadway Signs 288
9.3   Access Management 289
9.3.1   Interstate Access (Interchange) 290
9.3.2   Access Spacing 291
9.3.3   Driveway Spacing 291
9.3.4   Median Treatments 292
9.3.5   Exclusive Left- and Right-Turning Lanes 292
9.4   Intelligent Transportation System 293
9.5   Traffic Incident Management 293
9.5.1   Incident Management Concepts 293
9.5.2   Road Assistance Patrol Services 295
9.5.3   Traffic Incident Management Performance Measurement ...295
9.6   Transportation Demand Management 297
9.6.1   Operational-Based TDM Strategies 297
9.6.1.1   Tolling-Based TDM 297
9.6.1.2   High-Occupancy Vehicle Lane Strategy 297
9.6.1.3   High-Occupancy Toll Lane Approach 298
9.6.1.4   Variable Tolling Lanes (VTL) Practice 298
9.6.1.5   Cordoned Area Tolling 299
9.6.2   Toll Collection 299
9.6.2.1   Vehicle Identification 300
9.6.2.2   Vehicle Classification 300
9.6.2.3   Toll Processing 301
9.6.2.4   Toll Enforcement 301
9.7   Traffic Congestion Performance Measures 301
9.7.1   Average Speed 302
9.7.2   Roadway Level of Service (LOS) 302
9.7.3   Travel Time Index (TTI) 302
9.7.4   Average Commute Time 302
9.8   Spot Speed 302
9.8.1   Measuring Methods 303
9.8.1.1   Automatic Road Tube 303
9.8.1.2   Radar Meter Speed Measurement Method 304
9.8.1.3   Manual Stopwatch Method 304
9.8.2   Descriptive Statistics of Speed Data 306
9.8.2.1   Average Speed 306
9.8.2.2   Median Speed 306

9.8.2.3   Modal Speed 307
9.8.2.4   Standard Deviation 307
9.8.2.5   Percentile Speed 307
9.9   Next Step 308
9.10   Summary 308
9.11   Discussion 309
9.12   Exercises 309
Bibliography 312

10   **Maintenance**...**313**
10.1   Roadside Vegetation Management 313
        10.1.1 Roles of Roadside Vegetation 314
        10.1.2 Mowing 315
                10.1.2.1 Mowing Frequency 315
                10.1.2.2 Type of Mowing 316
                10.1.2.3 Mowing Safety 316
        10.1.3 Herbicide Application 318
                10.1.3.1 Nonselective Herbicides 318
                10.1.3.2 Selective Herbicides 318
10.2   Winter Snow and Ice Control 318
        10.2.1 Goal of Snow and Ice Control 319
        10.2.2 Mechanism for Snow and Ice Control................320
                10.2.2.1 Snow Removal 320
                10.2.2.2 Preventing Ice Formation 320
                10.2.2.3 Deicing or Preventing Ice Bonding to
                          Roadway Surface 320
        10.2.3 Road Weather 321
        10.2.4 Road Salt 321
        10.2.5 Other Effects from Road Salt 322
10.3   Pavement Maintenance 323
        10.3.1 Pavement Management System 323
                10.3.1.1 Pavement Reconstruction 324
                10.3.1.2 Pavement Rehabilitation 324
                10.3.1.3 Pavement Maintenance or Corrective
                          Maintenance 324
        10.3.2 Flexible Pavement Failures 325
                10.3.2.1 Flexible Pavement Cracking 325
                10.3.2.2 Pavement Deformation 328
                10.3.2.3 Material Failure 329
                10.3.2.4 Degradation 329
        10.3.3 Flexible Pavement Pothole Repairs, Crack Sealing
                and Filling 330

10.3.4 Rigid Pavement Failure 331
    10.3.4.1 Linear Cracking 331
    10.3.4.2 Durability (D) Cracking 332
    10.3.4.3 Shrinkage Crack 332
    10.3.4.4 Warping Crack 332
    10.3.4.5 Corner Breaks 333
    10.3.4.6 Scaling 333
    10.3.4.7 Pumping 334
    10.3.4.8 Faulting 334
    10.3.4.9 Spalling 334
    10.3.4.10 Polished Aggregate 335
10.3.5 Rigid Pavement Repair 336
10.4 Transportation Asset Management 336
10.4.1 Basic Engineering Economic Principle 336
    10.4.1.1 Interest Rate 336
    10.4.1.2 Time Value of Money 338
10.4.2 Equivalence 340
10.5 Next Step 342
10.6 Summary 342
10.7 Discussions 342
10.8 Exercises 343
Bibliography 344

Index 345

Foreword

Having spent most of my career working on transportation planning and project development at the Maryland Department of Transportation, I can attest to the complex, lengthy, and often tortuous process involved in delivering major transportation projects in the United States. The process is based on a myriad of federal and state statutory and regulatory requirements that must all be fully met, or projects will not obtain necessary approvals to proceed forward, or if approved may be stopped through litigation by project opponents. It often takes professionals who are involved in planning and project development years to understand the entire life cycle process for projects and all the details involved in moving a project through each phase of that life cycle.

Connie Kelly Tang and Lei Zhang have provided a holistic coverage of the entire surface transportation project and program development process from the beginning of planning through environmental approval, design, right of way acquisition, construction to operations and maintenance. They have brought together in one textbook a comprehensive reference manual and guidebook that will enable students and professionals in the field to see the big picture, as well as the details involved in the development of a transportation project.

They begin with the legal foundation for the process by reviewing relevant federal statutory and regulatory requirements. They then go through a logical sequencing of topic by describing each of the steps involved in a project's life cycle.

With each project development phase, key players are identified; agency roles are specified; and deliverables are outlined. The authors show how each phase builds on the work and outputs of the previous phase and how the following phases are going to use deliverables from the current phase.

The authors provide information on engineering and scientific fundamentals that project development professionals need to know, as well as how these are translated into standards and specifications. Key concepts and ideas are supported by illustrative drawings, pictures, charts, maps, and real-world example projects.

This book should be a valuable text for entry-level courses on transportation, as well as a resource and reference manual for students involved in senior capstone or graduate school projects.

Transportation agencies and consulting firms often lament about how graduates who are starting their career in transportation do not understand what is involved in the development and delivery of surface transportation projects. These organizations often have to make major investments in teaching their new employees about the planning and project development process and requirements. This textbook will provide students, as well as entry-level professionals, with a comprehensive grounding that will give them a head start as they begin their careers. It will also serve as a reference manual for more seasoned professionals who specialize in a certain part of the process about earlier or later steps in the overall project development process.

I only wish that I could have had such a comprehensive overview of transportation project and program development when I started my career at the Maryland Department of Transportation. It would have enabled me to be more effective in taking on leadership roles in delivering the transportation program that improves the economy and quality of life of the citizens and travelers in the State of Maryland.

Neil Pedersen
Executive Director
Transportation Research Board
National Academies of Sciences, Engineering, and Medicine
Washington, DC

Preface

Transportation program and project development is complex. The process spans over planning, programming, environment, design, right of way, construction, operations, and maintenance. Professionals from (1) civil engineering (e.g., transportation, structural, geotechnical, hydraulic, and construction engineering), (2) planning (e.g., urban and regional planning, financial planning, community planning, urban design), (3) social and environmental sciences (e.g., air quality and noise, stormwater management, environmental design, wildlife and habitat, wetlands and farmlands, social science and vulnerable populations, historical and archaeological studies), and (4) business, project management, and data science, work together in a team relay to transform an idea into a highway, a transit hub, an airport, or a water facility.

It is challenging for any one person to master all the knowledge and skills needed to perform every relevant task. However, it is critical for all involved to understand how this "relay" works and how the societal, environmental, governmental, and regulatory contexts influence the process and the technical solution. Professionals who understand the process and see the big picture are those who rise to the top as leaders.

This book provides holistic coverage on the technical subject matter, processes and procedures, and policy and guidance associated with transportation project and program development. For each phase of the process, key products delivered, processes used, governing principles, foundations of applicable science and engineering, technologies deployed, and knowledge required are discussed. While all coverages reflect the practices of the United States, the logic, principles, science, and engineering are applicable to all countries of the world.

This book can be used in several ways. First, it can serve as an introductory textbook for undergraduate students by offering a full breadth of civil and environmental engineering, project management, and planning knowledge and its applications to the transportation arena. Second, students in capstone design courses will find the book beneficial as a reference and for multidisciplinary collaboration. Third, this book can be used as a textbook or reference for a graduate-level course in civil engineering, transportation engineering, planning, and project management. Lastly, this book can serve as a reference for professionals in the transportation arena, including management, planning, environmental studies, engineering, right of way acquisition, construction, operations, and maintenance.

The authors are thankful for all valuable inputs from their university colleagues. The authors also would like to express their gratitude to numerous professionals at the U.S. Department of Transportation, state Transportation Departments throughout the nation, Metropolitan Planning Organizations large and small, and private consulting businesses for their willingness to share their first-hand experiences and valuable insights. It is these short and long conversations and discussions selfishly provided by all the professionals enabled the authors to grasp all issues involved. For that, the authors are forever indebted to them.

Connie Kelly Tang
Lei Zhang

Authors

Connie Kelly Tang is a graduate of New York University, Abu Dhabi, with a B.A. in Economics. She also holds a master's degree in Applied Economics from the University of Maryland. Since graduation, she has provided technical assistance to a wide range of activities and programs in the transportation field, including her near-three years of service as an on-site consultant to the Office of the Secretary, the U.S. Department of Transportation. She has worked with wide range of professionals and leaders in private businesses, local, state, and federal agencies, and academic institutions regarding issues related to transportation program and project development, gaining fundamental and practical knowledge on transportation project development. Currently, Connie Tang is the assistant director for Research and Outreach at the Maryland Transportation Institute, University of Maryland.

Lei Zhang is the Herbert Rabin Distinguished Professor and director of the Maryland Transportation Institute in the Department of Civil and Environmental Engineering at the University of Maryland. He earned his Ph.D. in Transportation Engineering, a M.S. in Applied Economics, and a M.S. in Civil Engineering, from the University of Minnesota. Dr. Zhang obtained his B.S. in Civil Engineering from Tsinghua University. He teaches both graduate and undergraduate courses in areas of transportation engineering and performs advanced research in areas of transportation systems analysis; transportation planning; traffic control and management, transportation economics and policy; mathematical, statistical, and agent-based modelling and simulation supporting urban mobility and sustainability analysis.

Chapter 1

The Law and the Process

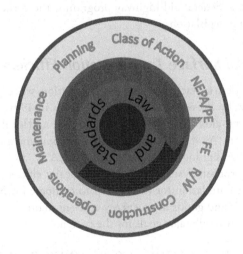

Modern highways link villages, towns, cities, mines, factories, farms, ports, markets, and, most importantly, people. This interconnected system not only enables the transportation of goods that sustain our basic physiological needs but also provides the means to satisfy the human desire to explore and discover. The highway system has become an integral part of our modern life.

The establishment, operation, maintenance, and planning for such a system require extensive human resources, materials, and funding capital. Governments are the ideal candidates to bear such duties and responsibilities. Throughout the world, governments have indeed taken on the responsibilities of planning, financing, operating, and maintaining their nation's roadway systems. Along the way, laws and regulations are enacted to ensure such complex systems function properly.

1.1 The Law

In the United States, the Federal government provides national policy and financial assistance to State and local governments to carry out the financing, planning, design, construction, opera-

> The U.S. surface transportation legislation provides authority, responsibility, and resource for surface transportation programs.

tion, and maintenance of the nation's roadway systems through the "Federal-aid highway" program. The hallmark of the Federal-aid highway program is that State and local government agencies, such as State, county, and city transportation departments, are empowered to make their own decisions concerning what projects are to be carried out. And by following Federal policies and guidelines, State and local governments receive federal aid funds and technical and policy support for their projects. The Federal-aid highway program is the cornerstone of the U.S. surface transportation legislation.

1.1.1 Intermodal Surface Transportation Efficiency Act (ISTEA)

ISTEA, the first modern surface transportation legislation, was enacted by Congress and signed into law by President George H. W. Bush in 1991. A total exceeding $155 billion in Federal-aid budget was allocated to State and local governments over the course of six years. As stated in President Bush's Statement of Administration Policy during ISTEA signing, the purpose of ISTEA is "to develop a National Intermodal Transportation System that is economically efficient, environmentally sound, provides the foundation for the Nation to compete in the global economy and will move people and goods in an energy efficient manner."

ISTEA outlined several major goals for the nation:

■ **Designating a National Highway System (NHS) Roadway Network**
  The act called for the designation of roadways that are the most important to interstate travel and national defense, connect with other modes of transportation, and are essential for international commerce.
■ **Returning Decision Making to State and Local Governments**
  The act gave State and local governments more flexibility in determining transportation solutions. The act officially created a new type of planning organization called the Metropolitan Planning Organization (MPO) through federal funding.
■ **Developing New Technologies**
  The act funded intelligent vehicle highway systems and prototype magnetic levitation system programs to develop new approaches for providing 21st-century transportation.
■ **Empowering Private Business Involvement**
  The act relaxed restrictions on the use of Federal-aid funds for toll roads. The private sector was explored as a source for funding transportation improvements.

■ **Funding Mass Transit**

The act continued discretionary and formula funds for mass transit.

■ **Enhancing Environments**

The act made activities that enhanced the environment as related to transportation projects (wetland banking, mitigation of damage to wildlife habitat, activities that led to meeting air quality standards, bicycle, and pedestrian projects, and highway beautification) eligible for Federal funding.

■ **Improving Highway Safety**

The act created a federally funded new program encouraging the use of safety belts and motorcycle helmets.

■ **Standardizing Vehicle Registration and Fuel Tax Reporting**

The act required states to ease the recordkeeping and reporting burden on businesses in order to increase productivity in the truck and bus industry.

ISTEA mandated that the U.S. Department of Transportation (DOT) develop and implement various programs through the Federal rulemaking process. As a result of the act, the Federal Highway Administration (FHWA) and the Federal Transit Administration (FTA) promulgated additional requirements, regulation, and guidance contained in Titles 23 and 49 of the Code of Federal Regulation (CFR).

ISTEA prescribed detailed procedural requirements on how state governments and local agencies should organize themselves and conduct their business in establishing transportation projects and programs. It also provided comprehensive guidelines and procedures covering virtually every aspect of the surface transportation project and program development as related to Federal-aid.

1.1.2 Transportation Equity Act for the 21st Century (TEA-21)

As ISTEA ended, on June 9, 1998, President William Jefferson "Bill" Clinton signed the Transportation Equity Act for the 21st Century (TEA-21) into law. TEA-21 authorized the Federal surface transportation programs for highways, highway safety, and transit for six years (1998–2003) with a budget of $218 billion.

Some key features of TEA-21 included:

■ Guaranteed Federal fund levels from the fiscal year 1998 to 2003
■ An extended Disadvantaged Business Enterprises (DBE) program with a flexible national 10% goal of DBE participation
■ Strengthened safety programs by promoting the usage of safety belts and enforcement of 0.08% blood alcohol concentration for drunk driving determination
■ Investment in research to maximize transportation system performance with special considerations for the intelligent transportation system.

1.1.3 Safe, Accountable, Flexible, Efficient Transportation Equity Act: A Legacy for Users (SAFETEA-LU)

Following the expiration of TEA-21, President George W. Bush signed the Safe, Accountable, Flexible, and Efficient Transportation Equity Act: A Legacy for Users (SAFETEA-LU) into law on August 10, 2005.

SAFETEA-LU continued many key programs of ISTEA but placed significant emphasis on multimodal transportation. The bill guaranteed funding totaling $244.1 billion for highways, highway safety, and public transportation for five years, starting in FY 2005 and ending in FY 2009.

Just as TEA-21, SAFETEA-LU covered virtually everything as related to surface transportation. However, SAFETEA-LU did reorganize and recompile various Federal-aid subjects into different titles, divisions, parts, chapters, sections, subsections, and sub-subsections.

Table 1.1a illustrated one of the core components of SAFETEA-LU, which is Title 1 – Federal-aid highways by various subprograms with authorized funding outlined. The Federal-aid program to states is divided into specific project and program categories. These categories include the Interstate Maintenance Program, the Bridge program, the NHS program, the Surface transportation program, the Congestion Mitigation and Air Quality (CMAQ) program, etc.

Table 1.1b SAFETEA-LU Title 5 Research Authorization Fund ($) is a budget summary under Title 5. The budget outlined under Title 5 in SATETEA-LU provided research funding to the transportation community. Budget information like these provides guidance and directions in priority setting and how to deal with changes.

Table 1.1a   SAFETEA-LU Title 1 Federal-Aid ($) Authorization Table

Section	Title I – Federal-Aid Description	Total $
1101(a)(1)	Interstate Maintenance Program (Sec. 119 & 104(b)(4))	25,201,594,690
1101(a)(2)	National Highway System (Sec. 103 & 104(b)(1))	30,541,832,710
1101(a)(3)	Bridge Program (Sec. 144)	21,607,441,525
1101(a)(4)	Surface Transportation Program (Sec. 133 & 104(b)(3))	32,549,756,505
1101(a)(5)	Congestion Mitigation and Air Quality Improvement Program	8,609,099,956
1101(a)(6)	Highway Safety Improvement Program (Sec. 148 & 104(b)(5))	5,063,922,785

(Continued)

**Table 1.1a (*Continued*)   SAFETEA-LU Title 1 Federal-Aid ($)
Authorization Table**

Section	Title I – Federal-Aid Description	Total $
1101(a)(7)	Appalachian Development Highway System	2,350,000,000
1101(a)(8)	Recreational Trails Program (Sec. 206)	370,000,000
1101(a)(9)(A)	Indian Reservation Roads (Sec. 204)	1,860,000,000
1101(a)(9)(B)	Park Roads and Parkways (Sec. 204)	1,050,000,000
1101(a)(9)(C)	Refuge Roads (Sec. 204)	145,000,000
1101(a)(9)(D)	Public Lands Highways (Sec. 204)	1,410,000,000
1101(a)(10)	National Corridor Infrastructure Improvement Program (Sec. 1302 of Act)	1,948,000,000
1101(a)(11)	Coordinated Border Infrastructure Program (Sec. 1303 of Act)	833,000,000
1101(a)(12)	National Scenic Byways Program (Sec. 162)	175,000,000
1101(a)(13)	Construction of Ferry Boats and Ferry Terminal Facilities (Sec. 147)	285,000,000
1101(a)(14)	Puerto Rico Highway Program (Sec. 165)	665,000,000
1101(a)(15)	Projects of National and Regional Significance (Sec. 1301 of Act)	1,779,000,000
1101(a)(16)	High Priority Projects Program (Sec. 117 and Sec. 1702 of Act)	14,832,000,000
1101(a)(17)	Safe Routes to School Program (Sec. 1404 of Act)	612,000,000
1101(a)(18)	Deployment of Magnetic Levitation Transportation Projects (Sec. 1307)	90,000,000
1101(a)(19)	National Corridor Planning and Development and Coordinated Border Infrastructure Programs	140,000,000
1101(a)(20)	Highways for LIFE (Sec. 1502 of Act)	75,000,000
1101(a)(21)	Highway Use Tax Evasion Projects (Sec. 143)(Sec. 1115)	127,100,000
1103(a)	Administrative Expenses (Sec. 104(a))	1,944,900,180

(Continued)

Table 1.1a (*Continued*)   **SAFETEA-LU Title 1 Federal-Aid ($)**
Authorization Table

Section	Title I – Federal-Aid Description	Total $
1103(f)	Operation Lifesaver	2,240,000
1103(f)	Rail-Highway X-ing Hazard Elim. in High Speed Rail Corridors	44,750,000
1104	Equity Bonus Program (Sec. 105)	43,664,356,609
1105	Revenue Aligned Budget Authority (Sec.110)	1,545,721,085
23 U.S.C. 125	Emergency Relief	500,000,000
1117	Transportation, Community, and System Preservation Program	413,031,303
1119(g)	Indian Reservation Road Bridges (Sec. 202)	70,000,000
1305	Truck Parking Facilities	25,000,000
1306	Freight Intermodal Distribution Pilot Grant Program	30,000,000
1308	Delta Region Transportation Development Program	40,000,000
1403	Toll Facilities Workplace Safety Study	500,000
1406	Safety Incentive Grants for Use of Seat Belts (Sec. 157)	112,000,000
1407	Safety Incentives to Prevent Operation of Motor Vehicles by Intoxicated Persons	110,000,000
1409	Work Zone Safety Grants	20,000,000
1410	National Work Zone Safety Clearinghouse	4,000,000
1411(a)	Road Safety (Data and Public Awareness)	2,000,000
1411(b)	Bicycle and Pedestrian Safety Grants (Clearinghouse)	2,300,000
1601	Transportation Infrastructure Finance and Innovation Act Amendments	610,000,000
1604(a)	Value Pricing Pilot Program	59,000,000
1803	America's Byways Resource Center	13,500,000

(*Continued*)

Table 1.1a (*Continued*)   SAFETEA-LU Title 1 Federal-Aid ($) Authorization Table

Section	Title I – Federal-Aid Description	Total $
1804	National Historic Covered Bridge Preservation	40,000,000
1806	Additional Authorization of Contract Authority for States with Indian Reservations	9,000,000
1807	Nonmotorized Transportation Pilot Program	100,000,000
1906	Grant Program to Prohibit Racial Profiling	37,500,000
1907	Pavement Marking Systems Demonstration Projects in Alaska and Tennessee	4,000,000
1909(b)	National Surface Transportation Policy and Revenue Study Commission	2,800,000
1919	Road User Fees Field Test – Public Policy Center of University of Iowa	12,500,000
1923	Transportation Assets and Needs of the Delta Region – Study and Strategic Plan	1,000,000
1934	Transportation Projects (ssambn to make specified allocations)	2,555,236,000
1940	Going-to-the-Sun Road, Glacier National Park, Montana	49,999,998
1943	Great Lakes ITS Implementation	9,000,000
1944	Transportation Construction and Remediation, Ottawa, OK	10,000,000
1945	Infrastructure Awareness	2,950,000
1960	Denali Access System Program	60,000,000
1961	I-95/Contee Road Interchange Study	1,000,000
1962	Multimodal Facility Improvements	20,000,000
		204,448,033,346

Source: Courtesy of the U.S. Department of Transportation, Federal Highway Administration, www.fhwa.dot.gov.

Table 1.1b   SAFETEA-LU Title 5 Research Authorization Fund ($)

Section	Content Description	Total ($)
5101(a)(1)	**Surface Transportation Research Program**	**982,000,000**
5201(g)	Exploratory Advanced Research Program	70,000,000
5201(i)(2)	Long-Term Pavement Performance Program	50,600,000
5201(j)(2)	Seismic Research (Setaside) (Sec. 502(g))	12,500,000
5201(m)	Biobased Transportation Research	50,000,000
5202(a)(2)	Long-Term Bridge Performance Program	31,000,000
5202(b)(3)(A)	Innovative Bridge Research and Deployment	65,500,000
5202(b)(3)(B)	High-Performance Concrete Bridge Research and Deployment	16,500,000
5202(c)(2)	High-Performing Steel Bridge Research and Technology Transfer	16,400,000
5202(d)	Steel Bridge Testing	5,000,000
5203(b)(2)	Innovative Pavement Research and Deployment	90,500,000
5203(b)(1)	Research to Improve Asphalt Pavement	16,400,000
5203(b)(1)	Research to Improve Concrete Pavement	16,400,000
5203(b)(1)	Research to Improve Alternative Materials Used in Highways	16,400,000
5203(b)(1)	Research to Improve Aggregates Used in Highways on the NHS	9,800,000
5203(c)(2)	Safety Innovation Deployment Program	51,000,000
5203(e)	Demonstration Projects and Studies	
5203(e)(1)	Wood Composite Materials Demonstration Project	2,000,000
5203(e)(2)	Asphalt and Asphalt-Related Reclamation Research – South Dakota School of Mines	1,500,000
5203(e)(3)	Development and Deployment of Techniques to Prevent and Mitigate Alkali Silica Reactivity	9,800,000
5203(f)	Physical demonstrations of TFHRC Work on Ultra-High Performance Concrete with Ductility	2,500,000

(Continued)

Table 1.1b (*Continued*)   SAFETEA-LU Title 5 Research Authorization Fund ($)

Section	Content Description	Total ($)
5204(g)	Fundamental Properties of Asphalts and Modified Asphalts	21,000,000
5204(g)	Transportation, Economic, and Land Use System	5,000,000
5206(b)	International Highway Transportation Outreach Program	1,500,000
5207(b)	Surface Transportation-Environmental Cooperative Research Program	67,500,000
5209(b)	National Cooperative Freight Transportation Research Program	15,000,000
5210(c)	Future Strategic Highway Research Program	205,000,000
5309(d)(1)	Centers for Surface Transportation Excellence	15,000,000
5309(d)(2)(A)	Center for Environmental Excellence	5,000,000
5309(d)(2)(B)	Center for Excellence in Surface Trans Safety at VTech	3,000,000
5309(d)(2)(C)	Center for Excellence in Rural Safety at Hubert H. Humphrey Institute, Minnesota	3,500,000
5309(d)(2)(D)	Center for Excellence in Project Finance	3,500,000
5501(c)	Transportation Safety Information Management System Project (TSIMS)	2,000,000
5502(d)	Surface Transportation Congestion Solutions Research Program	36,000,000
5503(e)	Motor Carrier Efficiency Study (To FMCSA)	5,000,000
5504	Center for Transportation Advancement and Regional Development	2,500,000
5506(d)	Commercial Remote Sensing Products and Spatial Information Technologies	31,000,000
5511	Motorcycle Crash Causation Study	2,816,000
5512	Advanced Travel Forecasting Procedures Program (TRANSIMS)	10,500,000
5513(a)	Thermal Imaging	2,000,000

(*Continued*)

Table 1.1b (*Continued*) SAFETEA-LU Title 5 Research Authorization Fund ($)

Section	Content Description	Total ($)
5513(b)	Transportation Injury Research at Center for Transportation Injury Research at Calspan University	5,000,000
5513(c)	Argonne National Laboratory-Advanced Transportation Technology Center	16,000,000
5513(d)	Feasibility Study for Creation of System of Inland Ports and Distribution	500,000
5513(e)	Automobile Accident Injury Research – Forsyth Institute	2,000,000
5513(f)	Rural Transportation Research	1,000,000
5513(g)	Rural Transportation Research Initiative– UGPTI at NDSU	2,000,000
5513(h)	Hydrogen-Powered Transportation Research Initiative	3,000,000
5513(i)	Cold Region and Rural Transportation Research, Maintenance, and Operations	4,000,000
5513(j)	Advanced Vehicle Technology – University of Kansas	10,000,000
5516(k)	Flexible Pavement and Extending Life Cycle of Asphalt	30,000,000
5513(l)	Renewable Transportation Systems Research – University of Vermont	1,000,000
7131	Hazardous Materials Research Projects (Setaside)	5,000,000
5101(a)(2)	**Training and Education (Sec. 504 and Sec. 5211 of Act)**	133,500,000
5204(a)(2)	National Highway Institute (Setaside) (Sec. 504(a))	48,000,000
5204(c)	Local Technical Assistance Program	55,500,000
5204(d)(2)	Garrett A. Morgan Technology and Transportation Education	5,000,000
5204(f)	Transportation Education Development Pilot Program	7,500,000

(Continued)

Table 1.1b (Continued)   SAFETEA-LU Title 5 Research Authorization Fund ($)

Section	Content Description	Total ($)
5204(h)(2)	Freight Capacity Building Program	3,500,000
5204(i)	Eisenhower Transportation Fellowship Program	11,000,000
5502(d)	Technical Assistance and Training to Disseminate	3,000,000
5101(a)(3)	**Bureau of Transportation Statistics (Sec. 111 of Title 49)**	135,000,000
1801(e)	National Ferry Database (Setaside NTE)	2,000,000
5101(a)(4)	**University Transportation Research (Secs. 5505-5506 of Title 49)**	364,400,000
5401(b)	National University Transportation Centers	160,000,000
5402(b)	University Transportation Research	188,500,000
5402(a)	Management and Oversight of Centers	2,000,000
5101(a)(5)	**Intelligent Transportation Systems Research**	550,000,000
5211	Multistate Corridor Operations and Management	35,000,000
5302(a)	ITS Outreach	1,250,000
5308	Road Weather Research and Development	20,000,000
5507	Rural Interstate Corridor Communications Study	3,000,000
5101(a)(6)	**ITS Deployment (Secs. 5208-5209 of TEA-21)**	122,000,000
Total – Title V		2,286,900,000

Source: Courtesy of the U.S. Department of Transportation, Federal Highway Administration, HCFB-1, www.fhwa.dot.gov.

1.1.4 Moving Ahead for Progress in the 21st Century Act (MAP21)

The Moving Ahead for Progress in the 21st Century Act (MAP21) was signed into law by President Barack Obama on July 6, 2012, following several extensions (also known as continuing resolutions) of SAFETEA-LU. MAP21 provided over $105 billion for FY 2013 and 2014 for the U.S. surface transportation programs.

MAP-21 created a streamlined and performance-based surface transportation program while continuing key policy programs (e.g., Safety program, NHS program, CMAQ program) that had existed in previous surface transportation legislation. Its performance measurement and management requirement provided a new way of holding State and local governmental agencies accountable for Federal-aid highway programs. FHWA's performance management requirements under Title 23 CFR Section 490 are a direct requirement from MAP-21.

Table 1.2 titled "TEA21 Authorized Apportioned Federal-Aid to States" is a summary of the Federal-aid dollar amount apportioned to each of the 50 states and the District of Columbia. While the State of Californian received over $3.5 billion on an annual basis for each of the three fiscal years, the smallest State Delaware received just over $163 million per year. Funding authorized by Federal surface legislation such as MAP21 provides a guarantee to state governments on the amount of Federal funds each State would receive. Such assurance is highly significant in terms of ensuring program continuity and certainty. The guarantee ensures that the State's transportation planning is more reliable, targeted, and functional.

1.1.5 Fixing America's Surface Transportation (FAST) Act

On December 4, 2015, President Barak Obama signed the Fixing America's Surface Transportation (FAST) Act into law. FAST Act provides long-term funding for states' surface transportation infrastructure planning and investment. FAST Act maintains a wide range of established highway-related programs in MAP-21.

FAST Act authorizes $305 billion over FY 2016 through 2020. The fund is allocated to the five U.S. DOT operating administrations covering highway, motor vehicle safety, public transportation, motor carrier safety, hazardous materials safety, rail, and research, technology, and statistical programs.

Through its formula programs, FAST Act distributes roughly 92% of total authorized Federal-aid highway program funding to State transportation departments. The FAST Act also creates a new formula-based National Highway Freight Program (NHFP), funded at an average of $1.2 billion per year. A new discretionary fund – the Nationally Significant Freight and Highway Projects, was established with an annual average budget of over $900 million. Table 1.3 summarizes all funding provisions for the highway program.

For the entire five-year period, a total over $226 billion is authorized with an average annual authorization of $45 billion. Of the $226 billion, the FHWA is authorized to obligate nearly $222 billion over the five-year period and over $44 billion annually without any further legislative directive.

Key FAST Act Federal-aid programs to state DOTs include:

■ National Highway Performance Program (NHPP): $116,399,144,775 (total) and $29,099,786,194 (annual),
■ Surface Transportation Block Grant Program: $58,268,082,929 (total) and $11,293,488,459 (annual),

Table 1.2 TEA21 Authorized Apportioned Federal Aid to States.

State	Summary of Apportionment Funding		
	FY 2012	FY 2013	FY 2014
Alabama	735,000,187	731,655,008	732,263,043
Alaska	485,764,026	483,552,446	483,955,039
Arizona	708,821,718	705,594,084	706,182,063
Arkansas	501,582,060	499,299,076	499,714,166
California	3,555,709,890	3,539,510,678	3,542,468,412
Colorado	518,042,180	515,683,066	516,112,989
Connecticut	486,582,741	484,366,739	484,770,705
Delaware	163,878,244	163,132,019	163,267,961
District of Columbia	154,578,359	153,874,511	154,002,708
Florida	1,835,524,503	1,827,170,634	1,828,689,002
Georgia	1,250,897,118	1,245,202,087	1,246,238,772
Hawaii	163,854,386	163,108,318	163,244,192
Idaho	277,093,190	275,831,738	276,061,294
Illinois	1,377,360,681	1,371,088,298	1,372,231,384
Indiana	923,106,579	918,904,028	919,668,926
Iowa	466,165,511	464,043,730	474,345,450
Kansas	366,100,850	364,434,474	364,737,489
Kentucky	643,689,561	640,759,832	641,292,458
Louisiana	679,945,133	676,850,440	677,413,014
Maine	178,831,529	178,017,304	178,165,560
Maryland	580,553,891	577,909,846	580,007,300
Massachusetts	588,382,905	585,702,797	586,191,765
Michigan	1,020,006,135	1,015,361,470	1,016,207,628
Minnesota	631,725,420	628,849,381	629,372,872
Mississippi	468,548,689	466,416,067	466,803,812
Missouri	917,135,156	912,960,643	913,719,741

(*Continued*)

Table 1.2 (*Continued*)   TEA21 Authorized Apportioned Federal Aid to States.

State	Summary of Apportionment Funding		
	FY 2012	*FY 2013*	*FY 2014*
Montana	397,487,710	395,678,315	396,007,464
Nebraska	280,019,456	278,744,779	278,976,662
Nevada	351,782,586	350,180,474	350,472,546
New Hampshire	160,065,929	159,337,106	159,469,843
New Jersey	967,284,837	962,878,904	963,682,664
New Mexico	355,764,458	354,145,060	354,439,590
New York	1,626,144,228	1,618,736,780	1,620,088,460
North Carolina	1,008,519,966	1,003,928,569	1,006,630,450
North Dakota	240,517,490	239,422,559	239,621,802
Ohio	1,298,574,906	1,292,661,761	1,293,739,008
Oklahoma	614,415,897	611,619,448	612,127,810
Oregon	484,226,759	482,022,424	482,423,497
Pennsylvania	1,589,522,664	1,582,285,178	1,583,603,275
Rhode Island	211,870,935	210,906,358	211,081,927
South Carolina	608,224,674	605,456,365	646,306,850
South Dakota	273,208,231	271,964,472	272,190,802
Tennessee	818,653,968	814,927,118	815,605,297
Texas	3,057,217,523	3,043,299,205	3,331,596,800
Utah	312,116,746	310,695,897	335,148,600
Vermont	196,619,042	195,723,810	195,886,832
Virginia	985,851,355	981,362,913	982,180,040
Washington	656,750,705	653,760,515	654,304,963
West Virginia	423,374,189	421,447,021	421,797,542
Wisconsin	728,941,489	725,623,241	726,226,908
Wyoming	248,186,872	247,057,049	247,262,623
Total	37,574,223,257	37,403,144,035	37,798,000,000

Source: Courtesy of the U.S. Department of Transportation, Federal Highway Administration, HCFB-1, www.fhwa.dot.gov.

- Highway Safety Improvement Program: $11,585,393,509 (total) and $2,250,328,071 (annual),
- Railway-highway Crossings Program: $1,175,000,000 (total) and $227,500,000 (annual),
- Safety-related Programs: $17,500,000 (total) and $3,500,000 (annual),
- CMAQ Improvement Program: $12,022,732,534 (total) and $2,334,684,018(annual),
- Metropolitan Planning Program: $1,717,082,358 (total) and $332,604,550(annual), and
- National Highway Freight Program(NHFP): $6,246,586,977 (total) and $1,115,461,959 (annual).

Table 1.3   Fast Act Highway Program Authorization ($)

Program Description	FY 2016–2020 Total	FY 2016–2020 Average Annual
Federal-aid Highway Program (Apportioned)	207,431,523,082	40,137,652,500
National Highway Performance Program	116,399,144,775	29,099,786,194
Surface Transportation Block Grant Program	58,268,082,929	11,293,488,459
Highway Safety Improvement Program (HSIP)	11,585,393,509	2,250,328,071
Railway-Highway Crossings Program	1,175,000,000	227,500,000
Safety-related Programs	17,500,000	3,500,000
Congestion Mitigation and Air Quality Improvement Program	12,022,732,534	2,334,684,018
Metropolitan Planning Program	1,717,082,358	332,604,550
National Highway Freight Program	6,246,586,977	1,115,461,959
FHWA Administration Expenses	2,333,976,918	456,397,500
General Administration/ARC	2,213,976,918	432,397,500
On-the-Job Training	50,000,000	10,000,000
Disadvantaged Business Enterprises	50,000,000	10,000,000
Highway Use Tax Evasion Projects	20,000,000	4,000,000
Federal Lands & Tribal Transportation	5,500,000,000	1,062,500,000

(Continued)

Table 1.3 (*Continued*) Fast Act Highway Program Authorization ($)

Program Description	FY 2016–2020 Total	FY 2016–2020 Average Annual
Other Programs	8,897,000,000	1,696,000,000
TIFIA	1,435,000,000	275,000,000
Territorial and Puerto Rico Highway	1,000,000,000	200,000,000
Nationally Significant Freight and Highway Projects	4,500,000,000	825,000,000
Construction of Ferry Boats	400,000,000	80,000,000
Emergency Relief	500,000,000	100,000,000
Nationally Significant Federal Lands and Tribal Projects	500,000,000	100,000,000
Appalachian Regional Development Program	550,000,000	110,000,000
Regional Infrastructure Accelerator Demonstration Program	12,000,000	6,000,000
Transportation Research	2,089,500,000	416,000,000
Highway Research & Development	625,000,000	125,000,000
Technology & Innovation Deployment	337,000,000	67,250,000
Training and Education	120,000,000	24,000,000
Intelligent Transportation Systems (ITS)	500,000,000	100,000,000
University Transportation Centers	377,500,000	73,750,000
Bureau of Transportation Statistics	130,000,000	26,000,000
Grand Total Authorizations	226,252,000,000	43,768,550,000
Contract Authority Exempt		
Emergency Relief	500,000,000	100,000,000
National Highway Performance Program	3,195,000,000	639,000,000
Contract Authority Subject to Ob Limitation	221,495,000,000	42,813,550,000
Obligation Limitation	221,495,000,000	42,813,550,000

Source: Courtesy of the U.S. Department of Transportation, Federal Highway Administration, www.fhwa.dot.gov.

1.1.6 Law, Programs, and Projects

Surface transportation legislation creates different transportation programs with their own respective goals. Often these programs have appropriated Federal funds, Federal specifications, and Federal standards.

For example, the NHS program created by ISTEA identifies national significant roads with dedicated funding to ensure their efficient operation. The NHS program has been continued throughout all the follow-up legislation and is still active.

The FAST Act establishes a new NHFP (NHPN) to improve the efficient movement of freight on the National Highway Freight Network. Under the NHPN program, the FHWA apportions over $1 billion to States on an annual basis to facilitate the implementation of the NHPN program.

The FAST Act also continues MAP-21 NHPP with dedicated funding exceeding $22 billion on an annual basis covering the fiscal years from 2015 to 2020.

Under each of these programs, the leading federal agency such as the FHWA may establish more specific rules known as the Code of Federal Regulations (CFR). These CFRs provide further guidance to state and local agencies to implement a program authorized and established by legislation (also known as public law).

State and local governmental agencies, following both the public law and Federal rules and regulations, develop projects under each program. A project is a specific, actionable activity to improve the capacity, safety, and reliability of the transportation system.

1.2 Other Programs

1.2.1 Continuing Resolution

A continuing resolution (CR) is a piece of legislation enacted by the U.S. Congress to operate Federal governmental departments, agencies, or programs with an appropriated fund for a short period. A CR occurs when a current piece of legislation is about to expire, but Congress and the President fail to determine how to move forward upon the expiration of the current legislation. A CR gets passed and signed into law by the President as a temporary measure, giving Congress more time to come up with a permanent solution. A CR is introduced to continue the current legislation at either its current funding level or a reduced level. Occasionally, it may take several CRs before final permanent legislation is enacted.

1.2.2 The University Transportation Centers Program

The University Transportation Centers (UTC) program has been part of the Federal-aid highway program since 1987. The UTC program furthers the goals of transportation research, innovative solution development, workforce education, and professional training in the field of transportation. Two- and four-year colleges and universities in the United States can join as a consortium to compete for dedicated funding under each legislation for a given topic or task area. Winning consortia receive Federal-aid funds to carry out research and tech transfer outlined in the submitted proposals.

1.2.3 Business Opportunities with the U.S. Government

The U.S. Federal government solicits a wide range of products and services, from buying tables and chairs to performing advanced research. All departments and operating administrations, including the U.S. DOT, the FHWA, the FTA, etc., announce a significant number of requests and solicitations through the websites of www.sam.gov and www.grants.gov. Any organization or person seeking to do businesses with the Federal government, including research entities, may sign up through the above web links for potential opportunities.

1.2.4 Hierarchy of Laws and Regulations

Not all laws and regulations are created equal. When there is a conflict between two laws or statutes, one must be disqualified. The U.S. Constitution is supreme to all other laws, statutes, and regulations. Below the Constitution are public laws passed by the U.S. Congress also known as the United States Codes (USC). Acting upon the directive of the USC or directed by Congress, the Federal executive agencies establish Federal regulations known as the CFR.

For a State, the State's Constitution is supreme to all state laws and regulations. Federal laws and regulations are supreme to state laws and regulations. State regulations are supreme to County or City regulation and ordinance. The hierarchy of these laws and regulations is illustrated in Figure 1.1.

1.2.5 Federal Highway Standards

Title 23 USC 109 requires that all design standards for projects on the NHS be approved by the Secretary of the U.S. DOT in cooperation with State highway departments.

The FHWA works with State highway departments on a wide range of research activities through the American Association of State Highway and Transportation Officials (AASHTO). As AASHTO develops various design standards and specifications, the FHWA, through its rulemaking authority, adopts such standard as mandatory requirements for the NHS roads and bridges. These NHS standards and specifications are typically then adopted by State highway agencies through states' rulemaking process.

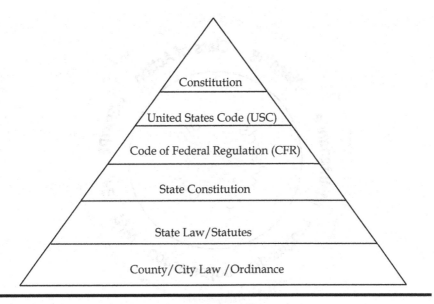

Figure 1.1 Hierarchy of laws and regulations illustration.

1.3 Transportation Project Development Process

Building a highway is a highly complicated task. Starting from the time when an idea is born, plans are drawn, right of way (ROW) is purchased, asphalt is laid, to cars, buses, and trucks traveling on it, a significant amount of money has been spent, and a wide range of professionals and the public have debated and resolved their differences.

To ensure a successful outcome, the current practice for transportation project development divides the entire process into different parts (phases), where professionals can utilize their specialized knowledge and expertise to offer the most optimal solution for a specific challenge.

1.3.1 Basics of the Development Process

All transportation projects and programs start with the planning phase. During this phase, three key issues are addressed. The first is an assessment of transportation demands or transportation needs; in other words, an analysis of what needs to be done both now (short term) and in the future (long-term) is performed.

The second is funding: what financial resources are available now and in the future. The last issue is to program and sequence individual projects with available funds in a time sequence that maximizes return.

During the planning stage, identifying what needs to be done is paramount. For example, should a new highway linking two cities be built? Should a two-lane

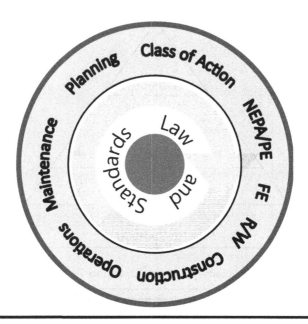

Figure 1.2   The circular path of the transportation project development process.

roadway be widened into four lanes to meet increased travel demand? Or should a bicycle lane be added to an existing roadway? These types of questions and issues are addressed in the planning phase. On the other hand, how to build a new highway, how to widen a roadway, or how to add a bicycle lane are not determined. The "how" to do things is left to other phases.

Following the planning, the next phase is the environmental impact evaluation (EIE). The EIE is where a planned project's real or potential environmental impact is screened and analyzed. The majority of highway projects have no impact on the environment. These projects can be screened out and moved directly to design and construction. For example, roadway resurfacing projects have virtually no impact on the environment and can be moved directly to the design and construction phase. Other projects, such as roadway widening, often require substantial amounts of environmental evaluation and analysis.

Environmental evaluation and analysis are carried out in concert with the preliminary engineering (PE) design, where a wide range of engineering solutions are compared. Through this EIE and PE process, a sound engineering design alternative that avoids, minimizes, and mitigates environmental impacts is achieved.

At the conclusion of the PE/EIE stage, an engineering design alternative with its key design location and concept is established (which accounts for approximately 10–30% of total design). Also, significant commitments may have been made to address public concerns (e.g., build a highway noise abatement wall) and the environmental regulatory agency requirements (e.g., wetland mitigation).

The next stage after the PE/EIE study is the final engineering (FE) design where a set of complete design plans is drawn, quantities of construction material are estimated and specified, and needed ROW (land) is mapped out. When the final design is complete, the so-called Plans, Specification, and Estimates package (PSE) is delivered, and most environmental permits are secured.

The ROW acquisition phases follow the final design phase. ROW agents can use the proposed ROW map to acquire the needed property. When all the needed ROW are acquired, the project moves to the construction phase through a process called open bidding. The winning company proceeds with construction while the owner of the project performs construction engineering inspection. Once a construction job is complete, the roadway opens to traffic.

As soon as a road opens to traffic, roadway signage may need adjustment; traffic signal timing may require optimization; new driveway connection and median opening requests are evaluated; traffic incidents need to be cleared. All these activities encompass an Operation department's primary task.

The maintenance department performs minor repairs to prevent minor issues from becoming major problems, ensuring that a roadway facility will last for as long as it is designed. Furthermore, maintenance monitors stormwater drains, mows the grass, removes snow, and clears debris.

Maintenance and operations are not the ends of the transportation project development cycle. Maintenance data and information are used by planning to assess the needs of the transportation system further.

1.3.2 Government Agencies and Organizations Involved

Given that roadway systems are considered part of public welfare, the administration of the roadway program has virtually always been part of government functions throughout the world, including the United States.

Under the authority of a wide range of Federal surface transportation legislation, the U.S. DOT administers the national highway program by partnering with State and local governments through its operating administrations – the FHWA, the FTA, the Federal Motor Carrier Safety Administration (FMCSA), and the National Highway Traffic Safety Administration (NHTSA).

The FHWA is mainly responsible for delivering the Federal-aid program covering both funding and technology as related to highway investment (Table 1.4). The FTA administers Federal aid to state and local transit agencies relevant to both public transportation fleet and fixed guideway facilities. The FMCSA is primarily responsible for the safe operations of commercial trucks and buses. The NHTSA develops and implements national vehicle safety standards.

State governments, through their state transportation or highway departments, carry out the actual day-to-day operations of all roadway related duties.

A State DOT's organizational structure is very similar to the FHWA HQ organizational structure, as shown in Figure 1.2.

Table 1.4   FHWA Washington DC Headquarter Program Offices

FHWA HQ PROGRAM OFFICES	Innovative Program Delivery	Serving as the agent for innovation through partnerships, technology deployment, and capacity building
	Federal Lands Highway	Manages Federal and Tribe land highway program (e.g., planning, design, construction)
	Policy and Governmental Affairs	Responsible for policy studies, data and information gathering, legislative analysis, and international program coordination
	Safety	Responsible for both safety technology and various safety programs
	Operations	Lead the transportation operations, management, and freight program
	Infrastructure	Responsible for asset management, pavements, construction, bridges/structures, program administration, and performance management
	Planning, Environment, and Realty	Responsible for planning, natural and human environmental analysis, guidance and methods, project development, environmental review, real estates
	Research, Development, and Technology	Lead the Exploratory Advanced Research and the comprehensive research covering Infrastructure, Operations, and Safety

Source: Based on Information from FHWA Homepage at www.fhwa.dot.gov.

Offices of roadway design, pavement design, stormwater design, hydraulic design, geotechnical design, structure (bridge and tunnel) design, environment, data and information, construction, operations, and maintenance are fundamental components of a state DOT.

Local governments, such as counties and cities, participate in transportation program and project development processes through local planning organizations called the MPOs. The MPOs are local transportation planning organizations authorized under Federal surface transportation legislation. An MPO's primary responsibility is to produce both short-term and long-term transportation plans.

Private businesses play a critical role in carrying out highway planning, environmental analysis, design, construction, operations, and research under contracts with various governmental agencies. Major highway construction work is always carried by using private businesses.

1.4 Professional Disciplines

> Professionals – people who have the technical, scientific, and engineering knowledge, skills, and abilities.

The complicated transportation project development process needs the participation of a broad range of professionals. At the planning stage, urban and transportation planning professionals, economists, civil or traffic engineers, data scientists, modelers, environmental scientists, and political and policy study majors, all play critical roles in assessing both the transportation and financial needs.

To fulfill the need of a thorough and credible environmental analysis during the EIE phase, professionals specializing in social studies, history studies, natural resource management, soil science, biological science, environmental sciences, and environmental engineering are all needed.

Professional licenses are often required to practice certain subject areas. Licenses are controlled by professional boards. State statutes sanction these boards under the auspices that such licensure protects the public welfare by maintaining a minimum standard and competence. In the absence of a government-sanctioned licensed profession, professional society or trade organizations may establish voluntary professional certification.

1.4.1 Civil Engineering Licensing

Under civil engineering, there are five subdisciplines. These five subdisciplines are (a) construction, (b) geotechnical, (c) structural, (d) water resource and environmental, and (e) transportation.

A person who is interested in being a licensed civil engineer in a state typically needs to have the following credentials.

- Graduating from an ABET (Accreditation Board for Engineering and Technology, Inc.) accredited engineering program with a B.S. degree (A person whose undergraduate degree is not engineering but with an engineering graduate degree may need to take remedial courses to qualify).
- Passing the comprehensive 5 hours and 20 minutes "Fundamentals of Engineering (FE)" exam administered by the National Council of Examiners for Engineering and Surveying (NCEES). Engineering undergraduates normally take this test during their senior year. Upon passing the test, the jurisdiction (State) issues the person a certificate known as Engineering in Training (EIT).
- Possessing adequate practical, qualified, and verifiable experience (typically four or more years).
- Passing a second "Principles and Practice" competency exam (administered by NCEES) divided into morning (AM) and afternoon (PM) sessions with

the PM session to be chosen from one of the five subject areas – (a) construction, (b) geotechnical, (c) structure, (d) water resource and environment, and (e) transportation.

Even though the civil engineering "Principles and Practices" exam has five subject areas to choose from, state licensing does not differentiate these specialties. A civil engineering license is a practice authority meaning that only a duly licensed person in a state may practice or offer to practice civil engineering.

Some states' statutes may also offer what is known as a "title" license and an "authority" license in addition to the general civil engineering practice license. "Title" authority means only these people who are licensed for that title can call themselves or advertise themselves under that title. For example, "Traffic Engineer" is a title license in the State of California. "Authority" license is typically a proficiency indicator license. The authority license shows that a person demonstrated a particular subject proficiency in a given area. Both title and authority licenses are subordinate to the primary practice licensing of civil engineering.

1.4.2 License Reciprocity or Comity

A person licensed in one state may be interested in practicing in another state. If that is the case, the person must get a new license from the other State. While basic requirements for licensing among most states are similar, there are still substantive differences. Fortunately, among most states, there is a reciprocity process where a licensed professional most likely will not need to retake all the examinations and can apply for a new license in a new state by following the reciprocity procedure.

1.4.3 Planning Professionals

There are no systematic governmental licensing schemes for planning professionals. The American Planning Association (APA) offers certification known as the American Institute of Certified Planners (AICP). The AICP certification (https://www.planning.org/certification/) indicates that the certified person meets APA's academic qualification requirements with relevant work experience and essential skills needed for planning work.

1.4.4 Environmental Science

On the environmental science front, there is no systematic government-wide licensing scheme. However, specialized training and passing a test on a given subject such as hazardous material handling, highway noise assessment, and jurisdictional wetland boundary delineation, etc., may be required to perform an environmental analysis work.

1.5 Summary

Federal surface transportation legislation provides national policy, the basis for regulations and guidelines, and financial resources. Specifically, the Federal surface transportation legislation provides Federal agencies and State and local governments with (a) the authority to develop needed transportation systems, (b) the duty and responsibility to deliver a transportation program to the public, and (c) the financial resource required to carry out the authority and mission.

The financial resource provided covers virtually every aspect of highway transportation program and project development. Transportation planning, environmental analysis, environmental impact mitigation, engineering design, ROW acquisition, construction, highway operations, and maintenance are activities eligible for Federal-aid highway fund.

Through the review, examination, and understanding of transportation legislation, professionals in the transportation arena will be able to gain a better understanding of new directions, new focuses, and new opportunities. Also, professionals will be able to prepare themselves adequately to meet new demands and challenges.

1.6 Discussion

1. What does Federal surface transportation legislation cover?
2. What is the hierarchy of U.S. laws and regulations?
3. What's the reason to become a licensed professional engineer?

1.7 Exercises

1. The U.S. Surface transportation legislation provides the legal authority, responsibility, and resources ($) to Federal, State, and local governments to develop various transportation systems. True or False
2. ISTEA authorized the creation of the National Highway System (NHS). True or False
3. ISTEA authorized the establishment of the Metropolitan Planning Organization (MPO). True or False
4. ISTEA provided state and local governments with more than $155 billion Federal-aid money. True or False
5. TEA-21 provided guaranteed funding to State and local governments. This guaranteed Federal funding provided great certainty to State DOT's projects. True or False
6. In Title I, Section 1961 of SAFETEA-LU, a project titled "I-95/Contee Road Interchange Study" was listed. Projects like this one are often reviewed as "pork barrel" projects meaning elected official appropriates Federal fund

for local specific projects solely or primarily to bring money to an elective's district. Do "pork barrel" projects bring more confidence to the political system? Yes or No

7. FAST Act has established a strong performance management program. True or False

8. When Congress and the President can't agree on how to extend expiring legislation, a Continuing Resolution is typically used to gain additional time to work out the differences. True or False

9. In the United States, Federal law is supreme to state and local law and regulation. True or False

10. To be a licensed professional civil engineer, a person needs to have the following credentials. Select all applicable ones.
    a. An ABET graduate
    b. Have a Master of Engineering graduate degree
    c. Pass the Fundamental of Engineering test
    d. Minimum four-year engineering experience
    e. Pass the Principles and Practice competency Exam
    f. Have a Ph.D. engineering graduate degree

11. A registered professional engineer in the State of California can practice engineering in the State of New York. True or False.

12. Federal agencies advertise and announce all their business opportunities on the websites of www.sam.gov or www.grants.gov. True or False.

13. An ABET-accredited civil engineering degree program is advantageous to graduates. True or False.

14. Gaining a professional certification is a waste of time. True or False.

15. Transportation projects and program development are complex. People who can see the big picture often become the leader in the industry. True or False.

Bibliography

ABET Accredited Program Search, ABET, http://main.abet.org/aps/accreditedprogram search.aspx.

Comparing Federal & State Courts, U.S. Courts, https://www.uscourts.gov/about-federal-courts/court-role-and-structure/comparing-federal-state-courts.

Court Role and Structure of Federal Courts, U.S. Courts, https://www.uscourts.gov/about-federal-courts/court-role-and-structure.

EIT and PE Engineering Licensure, National Council of Examiners for Engineering and Surveying, https://ncees.org/.

Fixing America's Surface Transportation (FAST) Act of 2015, Public Law 119-94, https://www.govinfo.gov/content/pkg/PLAW-114publ94/html/PLAW-114publ94.htm.

Get Certified, American Institute of Certified Planners, https://www.planning.org/certification/.

Intermodal Surface Transportation Efficiency Act (ISTEA) of 1991. Public Law 102-240, 105 Statutes 1914, https://www.govinfo.gov/content/pkg/STATUTE-105/pdf/STATUTE-105-Pg1914.pdf.

Moving Ahead for Progress in the 21st Century Act (MAP21) of 2012, Public Law 112-141, https://www.govinfo.gov/content/pkg/PLAW-112publ141/html/PLAW-112publ141.htm.

Safe, Accountable, Flexible, Efficient Transportation Equity Act: A Legacy for Users (SAFETEA-LU) of 2005, Public Law 109-59, https://www.govinfo.gov/content/pkg/PLAW-109publ59/pdf/PLAW-109publ59.pdf.

Transportation Equity Act for the 21st Century of 1998, Public Law 105-178, https://www.fhwa.dot.gov/tea21/tea21.pdf.

Chapter 2

Transportation Planning

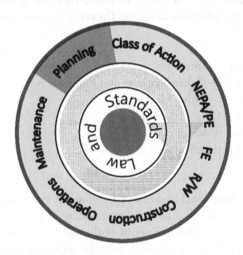

Transportation planning is one of the most fundamental steps in the transportation project and program development process. The planning process (a) identifies both short- and long-term transportation project and program needs, (b) estimates both project and program costs and available revenues, and (c) produces a work program (a project list) detailing when projects are scheduled to take place with appropriate funding identified.

2.1 Basic Concepts

2.1.1 Roadway Classification

2.1.1.1 Ownership-Based Approach

The most apparent roadway classification scheme is by ownership: who owns the road. A road can be either privately or publicly owned. Privately owned roads may or may not be open to the public. In the United States, it is more likely that roads are publicly owned. Different levels of Government, from Federal to State and local, all own roadways. Over 75% of public roadways are owned by local counties and townships.

Public roads are open to all traveling public. They can have a toll or be toll-free. Travel restrictions on public roads are applicable to all people without discrimination (Table 2.1a).

2.1.1.2 Functionality-Based Approach

A utility-oriented method for classifying roadways is through a road's intended functionality: what the road is designed and constructed to accomplish.

Using the Federal Highway Administration's (FHWA) Roadway Functional Classification Guidance, a road can be classified into one of the seven functional classes based on its intended functionality. The seven roadway functional classes are as follows:

1. Principal Arterial – Interstate
2. Principal Arterial – Other Freeways and Expressways
3. Principal Arterial – Other Highways
4. Minor Arterial – Other Highways
5. Major Collector Roadways
6. Minor Collector Roadways
7. Local Roads

The seven roadway functional classes can be grouped into three main categories: Arterial, Collector, and Local. The main function of an arterial roadway is to provide mobility with a high throughput of both vehicles and travelers with minimum to no impedance from factors such as traffic signals or driveway connections. An arterial roadway is typically free of access points. It connects with other roads through grade-separated interchanges (Table 2.1b).

The opposite end of an arterial roadway with respect to its functionality is the local road. Instead of focusing on throughput, local roads provide access points to abutting homes and businesses. Local roads link homes, businesses, and other establishments to higher functional class roads in a network.

In between arterial and local roadways, there are collector roadways. Collector roads gather traffic from local roadways and funnel them into arterials and conversely, distribute traffic from arterials to local roadways.

2.1.1.2.1 Principal Arterial – Interstate (I)

Interstate is the highest classification among all arterials. An Interstate highway is always a divided highway meaning travel in opposite directions is separated by either a grassy median or a physical barrier. Interstate highways are always free of

Table 2.1a U.S. Public Road Length (Mile) by Ownership

	State Highway Agency	County	Town, Township, Municipal	Other Jurisdictions	Federal Agency	Total
Rural	611,531	1,551,117	561,695	68,524	158,615	2,951,482
Urban	170,055	250,586	788,614	8,348	7,830	1,225,433
Total	781,587	1,801,703	1,350,308	76,872	166,445	4,176,915

Source: Courtesy of FHWA Highway Statistics 2020 HM16 data, www.fhwa.dot.gov.

Table 2.1b U.S. Public Road Length (Mile) by Functional Class

Area	Inter-State	Other Freeways Expressways	Other Arterial	Minor+Major Collector	Local	Total
Rural	29,280	6,504	223,907	667,649	2,024,142	2,951,481
Urban	19,160	12,100	178,921	146,936	868,317	1,225,434
Total	48,440	18,604	402,828	814,585	2,892,459	4,176,915

Source: Courtesy of FHWA Highway Statistics 2020 HM20 data, www.fhwa.dot.gov.

access points and connect to other roads through grade-separated interchanges. Interstate highways are designed and operated with mobility as the primary focus. The vast majority of Interstate highways have a posted speed limit of 55 mph or higher (Figure 2.1a).

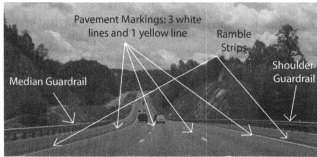

Pavement Markings: 3 white lines and 1 yellow line

Ramble Strips

Shoulder Guardrail

Median Guardrail

• Hilly terrain
• Physical median with median guard-rail
• Shoulder guardrail
• Three lanes per travel direction
• Solid yellow and white pavement marking lines

• Flat terrain
• Two lanes per travel direction
• Physical median – the depressed grassy median prevents storm water from flowing to the road
• Roadway signs

Rural Interstate Highway

Depressed Grassy Median

Urban Interstate

• Median - often is paved and has a physical barrier
• Median width - typically narrower than rural interstate
• Paved shoulder – typically wider than rural interstate
• Clear zone – substantially narrower than rural interstate
• Interchanges – more dense than rural interstate

Figure 2.1a   Interstate highway illustration.

2.1.1.2.2 Principal Arterial – Other Freeway and Expressway (OFE)

The design and operation of Other Freeway and Expressways (OFE) highways are very similar to those of Interstate highways. An OFE is always a divided highway with a grassy median or a physical median barrier, free of access points except for grade-separated interchanges. The vast majority of OFEs has a posted speed limit higher than or equal to 55 mph. Unlike Interstate highways, OFEs are located mainly in urban areas. Urban area beltways are classic examples of OFEs.

2.1.1.2.3 Principal Arterial – Other Principal Arterial (OPA)

The Other Principal Arterial (OPA) is designed with mobility as its main function. The main difference between OPAs and OFEs is its access controls. Unlike Interstate and OFE highways, OPAs may have at-grade-level access points where traffic lights are used to govern how traffic flows (Figure 2.1b).

2.1.1.2.4 Minor Arterial (MiA)

A Minor Arterial (MiA) is designed and operated with mobility as its key function. However, MiA highways have more at-grade connections with other roadways than OPAs. Additionally, a MiA highway is typically shorter in its overall length than an OPA highway. Posted speeds for MiA roadways usually do not exceed 45 mph.

2.1.1.2.5 Major Collector

Major collector roadways are designed and operated to offer both mobility and accesses. A major collector roadway collects traffic from local roads and funnels the traffic to an arterial. Conversely, a major collector roadway also distributes traffic from an arterial to local roads. A major collector roadway is generally situated within a small geographical area and does not cross through an entire region. Posted speed for major collectors is generally below 45 mph.

2.1.1.2.6 Minor Collector

Compared with major collectors, minor collectors have more access points. Minor collectors are always under local jurisdiction. A minor collector typically does not cross a large geographical area. Its overall length is generally short.

2.1.1.2.7 Local Roadway

Local Roadways are designed and operated with one critical functional goal: to provide connections to abutting homes and businesses. The length of a local road is typically short. It carries a low volume of traffic. Traffic on local roadways mainly comes from surrounding homes and establishments. Through traffic is a minimum

- Rural setting
- One lane per travel direction
- No physical median or median barrier
- Double yellow median lane marking lines with dashed pattern on the vehicle side permitting vehicles to pass on the opposite direction
- Paved shoulder
- Ditch and swale storm water system direction

- Rural setting
- One lane per travel direction
- No physical median or median barrier
- Double solid yellow marking lines prohibiting passing on the opposite direction
- No paved shoulder
- Ditch and swale storm water system direction

- Uran setting
- Three lanes per travel direction
- Physical raised median separator
- Sidewalk
- Curb and gutter swale storm water system direction.
- Utility poles

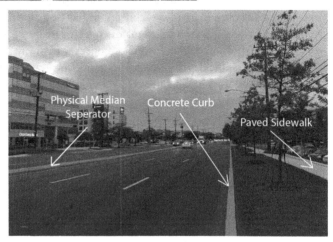

Figure 2.1b   Illustration of arterial roads.

- Urban setting
- One lane in one direction and 2 lanes in opposite direction
- One side with on-street parking
- Double solid yellow prohibiting passing on the opposite direction
- Dedicated bike lane
- Sidewalk
- Curb and gutter storm system

- Urban setting
- One-way street
- On street parking

- Rural and urban transition/mix zone
- Wide flush shoulder on one side of the road
- Curb and gutter on the other side of the road
- Paved shoulders function as bike lanes
- Center left lane offers storage space for left turning vehicles
- Curb and gutter drainage on one side and ditch swale on the other

Figure 2.1c   Illustration of collector roadways.

to none. Posted speeds for local roadways generally are no higher than 25 mph (Figure 2.1d).

2.1.1.3 Administrative Approach

The administrative approach classifies a roadway based on an administrative designation for the road. There are three administrative designations for a road, Strategic Defense Highway Network, National Highway System (NHS), and Truck National Network (NN). Administrative designation affects funding priorities for a roadway's design, construction, maintenance, and operations. A roadway may have an administrative designation in addition to its ownership and functional classification

Urban Local

• Urban setting
• Road is not divided
  – lack of pavement markings
• On street parking on both sides
• Individual drive-way connections

Urban Local

• Urban setting
• Local road/street - undivided and with no pavement markings
• On street parking on both sides.

Figure 2.1d   Illustration of local roads.

attributes. For example, a highway can be a State highway department-owned minor arterial and is designated as part of the NHS highway system.

Strategic Defense Highway Network – These roadways are critical to the strategic defense of the nation. The Strategic Defense Highway Network provides defense access, connectivity, and emergency capabilities.

National Highway System (NHS) – NHS roadways are public roadways most critical to the nation's economy, defense, and mobility. The NHS includes all Interstate highways, most principal arterial roadways, the entire Strategic Highway Defense Network, and some collectors and local roadways (Figure 2.1e).

Truck National Network (NN) – NN roadways are a subset of all public roadways meeting national minimum design and operating standards with regard to commercial truck operations as related to truck size and weight needs

Number of Highway Types Based on the FHWA Functional Classification Method: 7

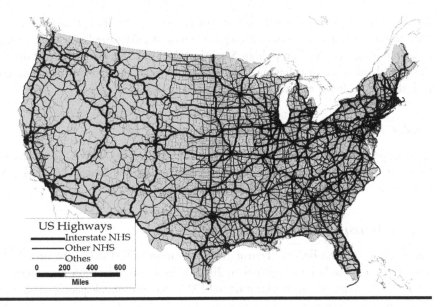

Figure 2.1e   Illustration of U.S. National Highway System highways.

2.1.2 Geographical and Statistical Reporting Area Designation

2.1.2.1 Urban Area

An urban area is a geographical location where human settlement is dense. A characteristic of an urban area is that the land is highly developed with housing, commercial buildings, roads, streets, and other man-made features.

The U.S. Census Bureau prescribes specific criteria to the designation of urban areas. The first type of urban area is the "Urbanized Area" where the area is inhabited by no less than 50,000 people. The second urban area type is the "Urban Cluster" where within the area, a minimum of 2,500 but less than 50,000 people reside.

2.1.2.2 Rural Areas

Rural Areas are all remaining areas that are not designated as either Urbanized Areas or Urban Clusters.

2.1.2.3 Metropolitan and Micropolitan Statistical Areas

For a host of statistical reporting purposes and Federal program implementation, the Office of Management and Budget (OMB) delineates Metropolitan (MSA) and Micropolitan Statistical Areas (MiSA). According to the OMB, a Metropolitan

Statistical Area shall have at least one urbanized area with adjacent areas having a high degree of social and economic integration with the urbanized area as measured by commuting ties. A Micropolitan Statistical Area shall have at least one urban cluster with adjacent areas having a high degree of social and economic integration with the urban cluster as measured by commuting ties.

2.1.2.4 Metropolitan Planning Area

A Metropolitan Planning Area (MPA), as defined by the FHWA, is an entire Urbanized Area plus contiguous areas expected to be urbanized in 20 years. An MPA is never smaller than its corresponding "Urbanized Area."

2.1.2.5 Transportation Management Area

The FHWA and the Federal Transit Administration (FTA) designate an urbanized area with populations higher than 200,000 as a Transportation Management Area (TMA). Transportation planning for TMAs has additional specifications and requirements.

2.1.3 National Air Quality Area Designation

An area's air quality designation directly affects the area's transportation planning requirements. The U.S. Environmental Protection Agency (EPA) carries out its legal responsibility under the Clean Air Act Amendment (CAAA) to monitor, classify, and designate an area regarding whether the area meets the National Ambient Air Quality Standards (NAAQS) for six common air pollutants. These six air pollutants are particulate matter (PM) (also called PM2.5 and PM10 or particle pollution), ground-level ozone (O_3), carbon monoxide (CO), sulfur dioxide (SO_2), nitrogen oxides (NO_2), and lead (Pb).

2.1.3.1 Non-Attainment Area

Based on field monitoring data, the EPA designates an area not meeting one or more of the six criteria pollutants under NAAQS as a non-attainment area (Figures 2.2a and 2b). An area classified as non-attainment must develop a State Implementation Plan (SIP) outlining steps to improve air quality. A SIP contains a series of actions needed to enable an area to achieve NAAQS standard. Transportation projects may be part of the actions needed to be acted upon to reduce emissions. For example, reducing new highway construction may reduce an area's vehicle travel demand, resulting in the reduction of vehicular-related air pollutants. A SIP is developed by a state environmental protection agency, not the state transportation department.

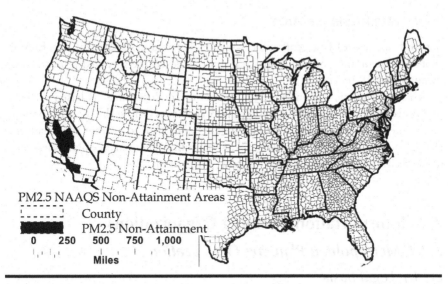

Figure 2.2a PM2.5 2012 NAAQS non-attainment area as of October 2015. (Drawn based on geospatial data courtesy of the EPA Green Book GIS Download, www.epa.gov.)

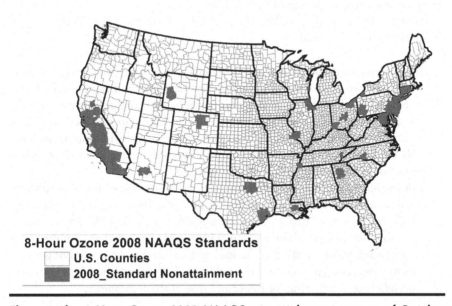

Figure 2.2b 8-Hour Ozone 2008 NAAQS non-attainment area as of October 2015. (Drawn based on geospatial data courtesy of the EPA Green Book GIS Download, www.epa.gov.)

2.1.3.2 Maintenance Area

A maintenance area is a geographical region that was designated by the EPA as "nonattainment area" originally but has since achieved the specified improvement through the SIP process.

This redesignation from "nonattainment" to "maintenance" is based on field monitored data showing no further NAAQS violations during the specified period. A "maintenance" area needs to have a maintenance plan to prevent the area from falling back into the "non-attainment" designation. A maintenance plan is less burdensome than the SIP for "nonattainment" designation.

2.2 Transportation Planning Organizations

2.2.1 Metropolitan Planning Organization

2.2.1.1 Legal Basis

ISTEA legislation mandates the official creation of a federally funded transportation policy-making organization called Metropolitan Planning Organization (MPO) for each urbanized area designated by the U.S. Census. Title 23 United States Code (USC) Section 134 outlines the specifics as related to the creation, operation, duty, and responsibilities of MPOs.

As stated in Section 123 USC, the establishment of the MPO Policy is that "it isin the national interest;

1. to encourage and promote the safe and efficient management, operation, and development of surface transportation systems that will serve the mobility needs of people and freight, foster economic growth and development within and between States and urbanized areas, and take into consideration resiliency needs while minimizing transportation-related fuel consumption and air pollution through metropolitan and statewide transportation planning processes identified in this chapter; and
2. to encourage the continued improvement and evolution of the metropolitan and statewide transportation planning processes by metropolitan planning organizations, State departments of transportation, and public transit operators as guided by the planning factors identified in subsection … Duties and responsibilities of an MPO are to lead and facilitate collaboration of governments, interested parties, and the public in developing a multimodal transportation plan reflecting the desire and goal of the community."

2.2.1.2 MPO Board

Any MPO operating in a TMA is required to be organized and run by a board.

An MPO board is the first component of the MPO entity. Its second component is its permanent technical operating division. The technical division performs all technical analyses and makes recommendations to the board for action.

An MPO's adopted bylaws govern the specific operating procedure of an MPO board in compliance with both Federal and state regulations.

The number of seats available in an MPO board is not set in any Federal statute. In cases where there are more local municipal entities than available seats (a board with too many members tends to be burdensome to run), an MPO Board can adopt bylaws to rotate seats among all municipal entities. Additionally, the chair of the board is typically set to rotate among members with an established term limit.

The MPO board shall consist of local elected officials, representatives of agencies operating alternative modes of transportation, and state officials.

Local elected officials can be city council members, mayors, county commissioners, county managers, or others.

Alternative modes of transportation representatives refer to the representatives from public transportation agencies operating transit rail and buses, public airport operators, and seaport operators.

State officials typically are the state department of transportation (DOT) representatives or gubernatorial appointees.

2.2.1.3 Voting Right

An MPO Board approves, adopts, and prioritizes projects and programs through voting. No Federal law prescribes who a voting member is, and no Federal law states how a vote should be counted.

An MPO Board can adopt its bylaw specifying who the voting members are and aren't. Also, its bylaw can specify how a vote is counted and establish a weighted voting system. No statute requires that all board members' votes be equal.

Board members are not MPO employees and do not receive a salary or any pay from the MPO board they are serving. Serving as an MPO board member is part of an elected official's duties and responsibilities. MPO board members meet periodically to discuss, debate, and approve or reject projects and programs.

The permanent technical operating division of an MPO consists of career technical professionals and is led by a technical director known as an Executive Director. Career technical professionals are MPO employees. These employees are responsible for carrying out all daily work and operations. Employees perform various technical analyses and make recommendations to the board for adoption.

2.2.1.4 Advisory Committees

An MPO Board can establish a wide range of advisory committees to make recommendations to the board and engage with the public as liaisons. Citizen's Advisory Committee (CAC) and Technical Advisory Committee (TAC) are two of the most

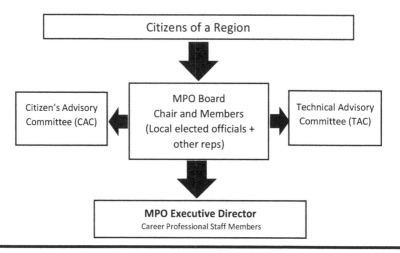

Figure 2.3   An illustration of a typical MPO organizational structure.

common ones established by an MPO board. While the CAC most often works directly with the board, providing input and feedback on issues, the TAC, consisting of professionals employed by relevant municipal agencies, works cooperatively with the MPO's professional technical staff.

Advisory committee members do not have the right to vote, nor are they paid by the MPO.

2.2.1.5 Professional Staff

Professional technical employees are hired to carry out the wishes and desires of an MPO Board under the management of the executive director. It is the technical director's responsibility to provide all logistical support for board activities. It is the technical staff's responsibility to develop preliminary projects and programs following policy guidance from the Board and make recommendations to the Board (Figure 2.3).

2.2.2 State Department of Transportation

All state transportation departments are established and authorized by state laws and statutes.

State transportation departments are multimodal. However, given rail transportation is mainly a private operation, airports and seaports are operated primarily by local governments, and aviation traffic is controlled by the Federal government, both the financial capital and human capital at a state DOT are heavily invested in the highway area.

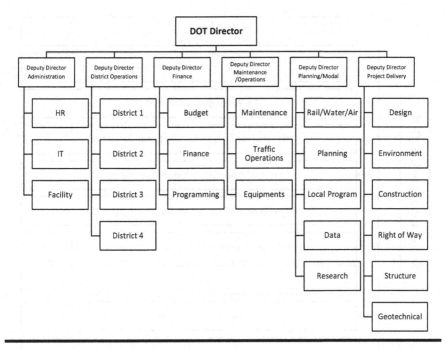

Figure 2.4   A typical organizational structure of a state DOT.

Figure 2.4 illustrates a typical state DOT structure.

With regard to highway planning, prior to the MPO era, state transportation departments were responsible for all roadway and highway planning. Nowadays, state transportation departments are primarily responsible for their statewide long-range transportation plans (LRTPs), statewide transportation improvement programs (STIPs), and the state's rural area transportation planning. For urban area planning, state highway agencies' role is mainly advising their local MPOs.

A state DOT is an executive agency of the state government. Its program and projects are typically approved by authorized leaders along the chain of command based on both Federal and state statutes and regulations. A DOT's approval process is different from the group approval used by an MPO Board.

2.2.3 The U.S. Department of Transportation

Within the U.S. DOT (Figure 2.5), the Federal Highway Administration and the Federal Transit Administration are responsible for carrying out Federal transportation planning regulations on highways and transit systems, as stated in 23 USC 134 and 135 and 23 CFR 450.

The FHWA and FTA provide both policy and technical advice to MPOs and State DOTs for their LRTP development and transportation improvement program

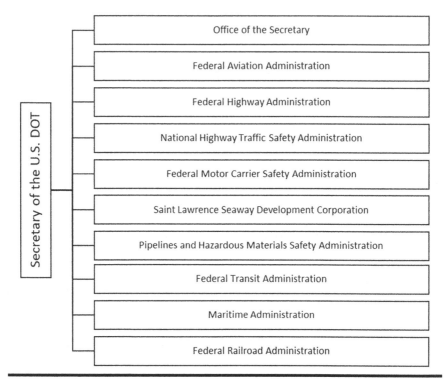

Figure 2.5 The U.S. DOT organizational structure. (Drawn based on informa-tion courtesy of the U.S. DOT, www.transportation.gov.)

(TIP). The FHWA and the FTA retain the final approval authority on the STIP and the transportation conformity determination.

Similar to state DOTs, the U.S. DOT is one of the U.S. Federal executive branches. Programs and projects are authorized and approved by authorized leaders along the chain of command based on Federal legislation and regulation.

2.3 Transportation Planning Product

2.3.1 Metropolitan Planning Organization Deliverable

By Federal law, MPOs are required to develop LRTPs consisting of goals, strategies, and projects covering no less than 20 years into the future. An MPO's long-range plan is typically referred to as a Metropolitan Transportation Plan (MPT).

An MPT is developed, maintained, and approved by the MPO board.

MPOs are also required to develop TIPs. A TIP consists of a list of trans-portation projects covering four years into the future. An MPO's TIP must be consistent with the MPO's MTP and should be updated at a minimum of every four years.

An MPO's TIP is approved by the MPO board and the state's governor.

2.3.2 State DOT Deliverable

State transportation departments are responsible for statewide LRTPs where statewide goals, strategic plans, and projects should be covered by no less than 20 years in the future. State DOTs develop their LRTPs and adopt them without the need for further approval by any third party.

For rural areas, state DOTs are required to cooperate with local officials associated with regional transportation planning organizations (if there is one) and local government agencies to develop both a rural LRTP and a rural TIP.

A State DOT is responsible for assembling all its respective MPOs' TIPs and rural TIPs to establish a statewide TIP. The STIP needs to be consistent with all Federal regulations.

A State DOT submits its STIP to FHWA and FTA for joint review and approval. The STIP is the only planning product requiring Federal approval.

2.3.3 U.S. DOT Deliverable

The U.S. DOT's FHWA and FTA do not generate project lists or establish project priorities for any MPO or any state DOT. The FHWA and FTA provide policy guidance to both state DOTs and MPOs in carrying out planning. The role of the FHWA and the FTA is to ensure that Federal planning regulations have adhered.

The FHWA and the FTA have the legal responsibility to review and approve statewide TIPs.

TRANSPORTATION IMPROVEMENT PROGRAM (TIP)

Four-year all-inclusive project list.

LONG-RANGE TRANSPORTATION PLAN (LRTP)

20-year outlook on goals, strategies, and projects.

Federal law and regulations also require that the FHWA and the FTA review an MPO's transportation planning process in TMAs no less than once every four years. This joint FHWA and FTA review is also known as an MPO certification review.

2.3.4 TIP Illustration

The New York Metropolitan Transportation Council (NYMTC) is the MPO for the New York City, Long Island, and the lower Hudson Valley. NYMTC is

Contents

Section I.	Preface	1
Section II.	Acronyms	2
Section III.	NYMTC and its Planning Area	3
	Membership	4
	Planning Area	4
Section IV.	Developing the TIP	5
	Documentation	5
	Building the TIP	5
	Required TIP Content	6
	Highway Safety Improvement Program (HSIP) and SPM	7
	Transit Asset Management (TAM)	10
	Infrastructure – Pavement and Bridge	18
	Transportation System Performance	21
	CMAQ Performance Plan	21
	Other System PM Targets	23
Section V.	Financing the FFYs 2020-2024 TIP	36
	Fiscal Constraint	36
	Operations and Maintenance	37
	Additional Funding Strategies	37
	Year of Expenditure	38
	Planning Factors	40
	Figures	41
Section VI.	Project Listings	50
	New York City	51
	Long Island	131
	Lower Hudson Valley	216
Section VII.	Appendices	308
	Appendix A	A-1
	Appendix B	B-1
	Appendix C	C-1
	Appendix D	D-1

Figure 2.6a   Sample TIP table of content from NYMTC. (Courtesy of New York New York Metropolitan Transportation Council at https://www.nymtc.org/.)

responsible for the region's transportation planning. Figure 2.6a shows the table of content of NYMTC's 2020–2024 TIP report.

Figure 2.6b shows the projects planned for Bronx County

2.4 Transportation Planning Procedure

While existing transportation infrastructure facilities enable a community's existence as it is, transportation planning reveals a community's vision for its future. To ensure that the transportation plan for a community reflects the community's value

** New York Metropolitan Transportation Council**

Bronx County Listing

AGENCY PIN WORKTYPE <AQ STATUS> / AQ CODE	PROJECT DESCRIPTION (COUNTY / TOTAL PROJECT COST)	FUND SOURCES & OBLIGATION DATE	TOTAL 5-YEAR PROGRAM (in millions of dollars)	PHASE	PRE FFY 2020	FFY 2020	FFY 2021	FFY 2022	FFY 2023	FFY 2024	POST FFY 2024
NYSDOT X73151 BRIDGE <Exempt>	BRUCKNER EXPRESSWAY (278I) OVERPASS REHABILITATION AT ROSEDALE AVE IN THE BRONX COUNTY TO REPAIR DETERIORATED BRIDGE ELEMENTS TO ASSURE CONTINUED SAFE OPERATIONS & A POSSIBLE FULL SUPERSTRUCTURE REPLACEMENT (BIN: 1075789)	STBG LG URB	0.000	PRELDES	1.640						
		SDF 08/2019	0.000	PRELDES	0.410	2.880					
		STBG LG URB 09/2020	2.880	DETLDES		0.720					
		SDF 09/2020	4.800	DETLDES					4.800		
		NHPP 11/2022	1.200	CONINSP					1.200		
		NHPP 11/2022	0.320	CONINSP					0.320		
		SDF 11/2022	0.080	CONST					0.080		
		SDF 11/2022	59.850	CONST					59.850		
AQC:A19	BRONX TPC:$70-$130M	TOTAL 5YR COST :	69.850		2.050	3.600	0.000	0.000	66.250	0.000	0.000
NYSDOT X73163 BRIDGE <Non-Exempt>	HUNTS POINT ACCESS IMPROVEMENT, PHASE 1 (EVERGREEN AVE TO BRUCKNER SHERIDAN INTERCHANGE) WILL RECONSTRUCT THE BRUCKNER EXPRESSWAY VIADUCT AND CONNECTING RAMPS TO EDGEWATER ROAD TO PROVIDE DIRECT HIGHWAY CONNECTIONS TO HUNTS PT MARKET	NHPP 02/2019	0.000	CONINSP	8.240						
		PIT BOND 02/2019	0.000	CONINSP	2.060						
		NHPP 02/2019	0.000	CONST	320.000						
		PIT BOND 02/2019	0.000	CONST	80.000						
		PIT BOND 02/2019	0.000	CONST	164.000						
		OTHER 02/2019	0.000	CONST	1.782						
		MTA 02/2019	0.000	CONST	0.016						
AQC:NON	BRONX TPC:$0.5-$0.7B	TOTAL 5YR COST :	0.000		576.098	0.000	0.000	0.000	0.000	0.000	0.000
NYSDOT X73164 BRIDGE <Non-Exempt>	HUNTS POINT ACCESS IMPROVEMENT, PHASE 2 (E 141 ST TO BARRETTO ST) WILL RECONSTRUCT & WIDEN THE BRUCKNER EXPRESSWAY VIADUCT AND CONNECTING RAMPS TO LEGGETT AVE TO PROVIDE DIRECT HIGHWAY CONNECTIONS TO HUNTS PT MARKET THEREBY REDUCING CONGESTION ON LOCAL STREETS AND IMPROVE QUALITY OF LIFE FOR THE SOUTH BRONX COMMUNITY (BINS: 1066669 & 1066666E)	PIT BOND 01/2020	10.400	CONINSP		10.400					
		PIT BOND 01/2020	655.200	CONST		655.20 0					
		PIT BOND 01/2020	10.400	DETLDES		10.400					
AQC:NON	BRONX TPC:$0.6-$0.8B	TOTAL 5YR COST :	676.000		0.000	676.000	0.000	0.000	0.000	0.000	0.000

Figure 2.6b Sample TIP project list adopted by the NYMTC. (Courtesy of New York Metropolitan Transportation Council at https://www.nymtc.org/.)

and vision, the process of developing the plan must be conducive to (a) public and community involvement, (b) business participation, and (c) professional leadership.

2.4.1 Planning Principle

Transportation planning relies on the 3C principle known as continuing, cooperative, and comprehensive planning. The application of the 3C principle is reflected in the organization and delivery of all planning products.

2.4.1.1 Public and Community Involvement

A transportation plan is not designed for the government but for the people. The test of its success is the public's thoughts and opinions of it. Consequently, steps and methods to get the public and the communities to participate in a meaningful and productive way are critical to its success.

THE 3CS FOR TRANSPORTATION PLANNING

Continuous, Cooperative, and Comprehensive Planning

2.4.1.2 Relevant Governmental Agencies, Organizations, and Businesses Involvement

There are a host of governmental agencies and offices involved in providing transportation services to the public. These agencies range from State, county, and city transportation and public works departments, to transit operators, airport operators, and seaport operators.

In addition to paying attention to public agencies, efforts to involve private businesses engaged in both passenger and freight transportation are highly desirable and necessary.

Bringing together all relevant agencies, organizations, and businesses ensures a comprehensive transportation plan.

2.4.1.3 Professional Leadership

To be successful, transportation planning relies on professionals with the necessary knowledge, skill, and ability to conduct comprehensive evaluations and sift through multiple options and alternatives.

2.4.2 Planning Steps

2.4.2.1 Inventory of Existing Facilities and Services

Performing an inventory of existing infrastructure facilities is the first step in transportation planning. Data items characterizing roadways infrastructure such as roadway geospatial location, layout, number of travel lanes, turning lanes, bicycle lanes, sidewalks, bus-bays, intersections, bridges, culverts, etc., should be part of the inventory. The geospatially enabled data offers not only precise location information but also time series and trending information.

2.4.2.2 Monitoring both Infrastructure Operating Condition and Performance

Data covering infrastructure inventory and condition (e.g., pavement condition, structure integrity (e.g., bridge rating), operating condition (e.g., traffic volume, speed, travel time reliability, number of crashes, and level of service), and performance characteristics (travel reliability, crash rate, etc.) are collected and processed to understand infrastructure conditions. Such data also offer foundational information to study causation relationships among various factors (e.g., how truck weight affects pavement deterioration, how speed relates to crash, how population growth affects travel demand).

2.4.2.3 Involving Public

Meaningful public and community involvement is critical for success. Any adopted processes and procedures designed to facilitate planning need to ensure that public input on moving forward receives adequate consideration. Public involvement needs to be a continuous effort throughout the planning process.

2.4.2.4 Gathering Land Use Data and Working with Local Zoning Officials

Land use refers to how a geographical area is used or will be used for various human activities (Figure 2.7). Types of activities that can be performed in a zone are governed by zoning regulations and laws. Typical land-use categories include single-family residential, high-density residential, light commercial, retail and business, office and retail, light manufacturing, heavy manufacturing, and industrial usage. Virtually all zoning regulations and laws are locally enacted (e.g., county and city governments pass local zoning ordinance).

Land use covers both current and planned land usage. While roadways affect land-use decisions, land-use decisions also significantly affect roadway needs. Roadway needs and land-use patterns are highly intertwined.

2.4.2.5 Gathering Social, Demographic, Economic, and Other Related Data

Data and information related to population, population growth, land use, land-use policy, economic growth, and population outlook are critical for quantitative analysis. The U.S. Census Bureau offers a significant amount of information on such data. To achieve quality data, data quality control procedures and protocols should

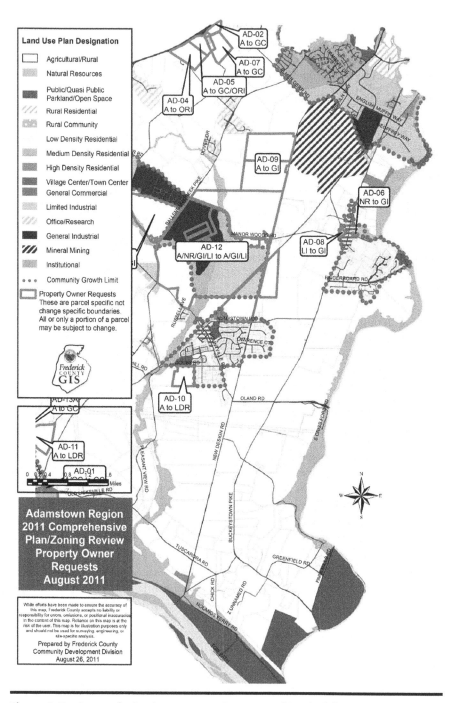

Figure 2.7 A sample land-use map. (Courtesy of Frederick County, Maryland Zoning GIS at https://frederickcountymd.gov/6779/Zoning-Map-Atlas.)

be established. Written procedure and written protocol with periodic updating are ways to ensure data quality.

2.4.2.6 Analyzing Transportation Demand through Modeling

Transportation infrastructure investment is costly. Before any real field investment is made (e.g., to reconstruct a road from two to four lanes or to construct new bike lanes on both sides of the road), the least costly method to test out a potential project's effectiveness is through computer modeling and simulation. Through computer modeling, a wide range of project options can be analyzed to identify their strengths and weaknesses.

2.4.2.7 Regional Air Quality Issue Consideration

For any area designated by the EPA as air quality non-attainment for PM (PM2.5/PM10), carbon monoxide (CO), and/or ground-level ozone (O_3), the transportation improvement plan for the area must be evaluated to meet air quality goals established in the area's State Improvement Plan (SIP).

Air quality impact analysis for a non-attainment area covers both the macro-level (all transportation projects) during the planning phase and the micro-level (individual project) during the project-level environmental analysis phase.

For air quality attainment areas, air quality analysis is only carried out at the project level (micro-level) during the environmental evaluation analysis phase for carbon monoxide.

> Regional air quality analysis as related to transportation planning is only needed for NAAQS non-attainment or maintenance areas.

For air quality non-attainment areas, macro-level analysis is to ensure that total vehicle emissions (e.g., volatile organic compounds, NO_x, PMs from all vehicles and all roads) from all transportation operations do not exceed a predetermined amount known as the SIP "emission budget" established for the mobile source.

2.4.2.8 Cost and Revenue Estimation

Cost estimates for a proposed action or project are typically based on historical data and experience. The availability of historical project and program cost data is vital to a planning agency's ability to perform cost estimation. One of the critical issues associated with project cost estimates is the need to specify the year the cost is based. For example, a project that costs 1 million 2001 dollars is different from 1 million 2020 dollars. Factors related to inflation and the time value of money must be considered given the cycle of the transportation project is often long.

In addition to cost estimation, both the source and quantity of revenue must be analyzed. The amount and sources of revenue for each of the four years covered in the TIP must be identified (e.g., for the year 2024, federal-aid fund: $350 million; State fund: $450 million; local fund: $125 million …).

Funding sources may range from local, state, and Federal government to other sources. Again, historical data on revenue is essential to the estimation. For each government entity, funding can come from fuel tax, vehicle tax, toll, sales tax, general revenue, and others.

Cost and revenue analysis is directly tied to the fiscal constraint requirement for a TIP per Federal regulation. Federal transportation planning regulation requires that the total cost of projects contained in a TIP must be constrained to available revenue. In addition to meeting the Federal law need, balancing expenditure and revenue is a standard business operation.

2.4.2.9 Assigning Appropriate Fund to a Project

During this step of the analysis, a proposed project is assigned with appropriate funding types. Given the wide range of revenue sources, certain revenue may have limitations or prohibitions on the types of projects or programs the fund can be used. The process of assigning funding is to optimize the overall fund usage.

2.4.2.10 LRTP and TIP Adoption

Throughout the steps outlined so far, a preliminary LRTP and a draft TIP are produced for presentation to the MPO policy board for further discussion, comment, modification, and approval or rejection.

The final LRTP and TIP adoption is decided through voting by Board members. Given that board members are elected political leaders, it is inevitable that the process is a political one. This is also why the final TIP and LRTP are often characterized as political products.

During an MPO Board's adoption processes, alternative solutions and projects may be proposed by board members. MPO staff analyze such alternative solutions and projects and report back to the Board for further discussion and adoption.

2.4.3 Planning Regulation

State DOTs and MPOs follow Federal transportation planning laws and regulations promulgated in 23 USC 134 and 135.

Fiscally Constrained TIP

A TIP must be fiscally constrained meaning projects identified in a TIP must have identifiable funding sources – revenue to cover their cost.

Conformity Determination

Applicable to air quality non-attainment areas only. A TIP must pass transportation conformity determination – meaning total vehicle emissions from all transportation projects (activities) within a non-attainment area cannot exceed a preestablished limit known as mobile source budget.

Critical components of the Federal transportation planning legislation are as follows:

■ Establishing and operating an MPO
■ Conducting meaningful public involvement
■ Cooperating with all agencies including private transportation operators in developing transportation plans
■ Ensuring that transportation improvement plans meet transportation conformity needs, and
■ Ensuring all projects contained in a transportation improvement plan are fiscally constrained, meaning the revenue or forecasted revenue must be enough to cover the projected cost.

2.5 Traffic Safety Data Analysis

Surface transportation projects have been developed to meet the needs for economic development, transportation capacity for congestion relief, or safety improvements singularly or in combinations. Traffic safety data analysis kicks off the development of the statewide Strategic Highway Safety Plan (SHSP), which is integrated into urban area TIPs and the Statewide STIP.

Three key datasets are collected and analyzed to perform safety assessments. These three datasets are crash data, roadway characteristic data, and traffic data.

2.5.1 Crash Data

Crash data refers to records when a vehicle collides with another vehicle, object, or structure and causes bodily injury or property damage exceeding a specific threshold limit. The property threshold limit varies among states throughout the United States but typically is around $1,000 per crash. Specific data items to be collected for these reportable crashes by law enforcement officers follow the so-called Model Minimum Uniform Crash Criteria (MUCC) developed by the National Highway Traffic Safety Administration (NHTSA) and the Governors Highway Safety Association (GHSA). The MUCC has over 70 data elements, which are divided into the following eight areas.

a. Crash data elements – crash type, date/time, location, first harmful event, crash/ collision impact manner, weather, light, roadway surface, crash severity, number of involved vehicles, number of motorists, number of non-motorists, fatality number, alcohol involvement, drug involvement, etc. There are 27 elements.

b. Vehicle data elements – vehicle identification number, make, model, year, registration state, body type, total occupants, travel direction, vehicle damage, the sequence of events, etc. There are 24 data elements.

c. Person data elements – name, date of birth, sex, injury status, seating position, restraint system, airbag deployment, election, license driver jurisdiction, driver action at the time of a crash, distraction, alcohol test, drug test, injury area, injury severity, etc. There are 25 elements.

d. Roadway data elements – bridges, roadway curvature, roadway functional class, access control, roadway lighting, grade, etc. A total of 13 elements are listed.

e. Fatal section – attempted avoidance maneuver, alcohol test, and drug test.

f. Large vehicle – trailer, cargo, etc. A total of 13 data elements are required.

g. Non-motorist data – non-motorist action prior to the crash, non-motorist safety equipment, etc. A total of six elements are required.

h. Dynamic data elements – motor vehicle automation and motor vehicle automated driving system.

2.5.2 Roadway Characteristic Data

Roadway characteristics data describe the physical attributes of all roadways systemically. While some of the roadway attributes are collected by law enforcement through the crash data report, systematic and more comprehensive roadway data enables a more thorough analysis of the entire system. FHWA recommends that States and local transportation agencies follow the Model Inventory of Roadway Elements (MIRE) as a standard to collect roadway characteristics data. Roadway data elements covered under MIRE are extensive, with a total of 202 data items. MIRE data are divided into three main categories: roadway segment, roadway alignment, and roadway junction.

The Roadway Segment Category is further divided into (a) segment location (e.g., segment length), (b) segment classification (e.g., functional class), (c) segment cross section (e.g., pavement roughness), (d) lane description (e.g., number of lanes, cross slope), (e) shoulder (e.g., curb presence, right shoulder type), (f) median description (e.g., median width), (g) segment roadside (e.g., clear zone width), (h) other segment descriptions (e.g., number of signalized intersections in segment), (i) segment traffic data (e.g., AADT), and (j) segment traffic operations/control data (e.g., speed limit).

The Roadway Alignment Category covers the horizontal curve (e.g., curvature, curve length) and vertical curve (e.g., percent of gradient and grade length) information.

The Roadway Junction Category is divided into at-grade intersections (e.g., intersecting angle) and grade-separated interchanges (e.g., ramp length).

2.5.3 Roadway Traffic Data

In addition to these traffic data collected per the MIRE guidance, State highway agencies also collect traffic data per FHWA's Highway Performance Management System (HPMS) Field Manual. Traffic volume data such as AADT are collected for all the NHS roadways.

2.5.4 Safety Data Analysis

Safety data analysis is carried out in various ways and approaches. During the planning phase, the goal is to discover safety or potential safety issues. Causes of safety concerns and solutions to mediate safety issues are typically addressed during engineering design and operations.

2.5.4.1 Total Crash Analysis

The first type of analysis is "total" analysis. The "total" analysis is a simple enumeration of all crashes of all types that occurred within a geographical area (e.g., a county) or a group of roadways (e.g., all Urban Interstate Highways in the State of New Jersey or associated with a road (e.g., I-95 in Baker County from MP 10 to MP 120) during a specific period (e.g., a calendar year).

$$T_{jk} = \sum_{k=1}^{n} (C_k)$$

Where
T_{jk} is the number of total crashes during period j covering crash type k (from 1 to n) for an area.
C_k it the total crashes for crash type k during the period j for an area.

Total enumeration data are often presented in a time series fashion to show trending. Figure 2.8a1 shows how total reported motor vehicle crashes for the entire United States occurred in a calendar year are analyzed. As the figure indicated, crashes experienced a slow but steady decline for more than a decade starting in 1996. However, the declining trend stopped in 2011. It has been a significant challenge for the nation since then as more crashes are occurring without abating. Notice the Y-axis: The magnitude is millions of crashes, which is very astonishing.

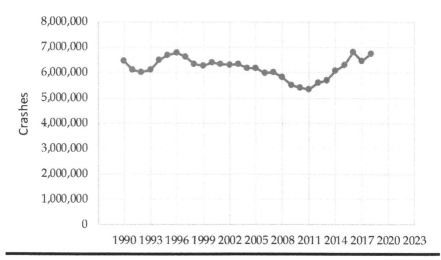

Figure 2.8a1   Total motor vehicle crashes in the United States. (Sources: Based on US DOT BTS National Transportation Statistics Table 2.17 at https://www.bts. gov/topics/national-transportation-statistics.)

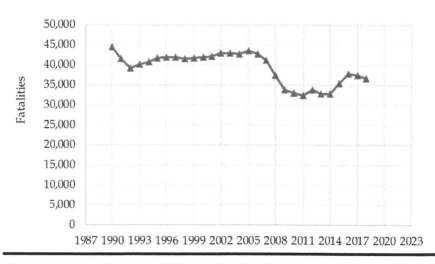

Figure 2.8a2   Total motor vehicle crash Related Fatalities in the United States. Sources: Based on US DOT BTS National Transportation Statistics Table 2.17 at https://www.bts.gov/topics/national-transportation-statistics

Figure 2.8a2 shows how fatalities that resulted from motor vehicle crashes for the entire United States are analyzed in the most fundamental way. Total deaths had been rising for an extended period starting from 1992. Since its peak in 2005, it had experienced a steep decline for nearly ten years. Unfortunately, the more recent trend has shown no further progress or minimum progress in reducing deaths.

2.5.4.2 Normalized Crash Data Analysis

The normalized crash data analysis refers to the derivation of crash rates by normalizing the total crash data. The crash rate typically has the units of crashes per 1 illion of the population in a region, crashes per 100 million vehicle miles traveled, and crashes per 1 million drivers.

The normalization needs denominator data, also known as exposure data. Exposure data often refer to the number of people in a region, licensed drivers, registered vehicles, and vehicle miles traveled in a region, etc. The crash rate is calculated by using the equation below.

$$R_{jk} = \frac{T_{jk}}{E_{jk}}$$

where R_{jk} is the crash rate for time j and crash type k.
T_{jk} is the total crashes occurred in time j and crash type k.
E_{jk} is the exposure data for time j and crash type k.

The key for normalization analysis is the gathering and computation of the exposure data for the corresponding area, roadways, and time. Normalized data and information can be used to contrast data among different regions, different periods, and various roads, enabling ranking and ordering.

The normalization data analysis is often used to carry out comparisons among different functional class roads and area type (e.g., rural vs. urban).

Figures 2.8b1 and b2 show how crash and fatality rates per 100 million vehicle miles traveled for the United States are constructed. These two normalized data charts, compared to the two total charts, give very different perspectives on crash and fatality trending. The desired results are that both the total and the rate are going down.

2.5.4.3 Geospatial Analysis and Visualization

Every crash has a geospatial location characterized by its latitude and longitude information in addition to other location information such as the municipality, county, and state where the crash occurred. Because of this data character, crash data can be analyzed geospatially, as illustrated in Figures 2.8c1–c4e–h.

Figure 2.8c1 shows the number of total fatalities in each state in 2019. A gradient color scheme, where the lighter the color is, the fewer fatalities occurred in that state, is used. The figure is also labeled with the total deaths in each state.

Figure 2.8c2 shows the number of 2019 fatalities per 100 million vehicle miles traveled. Again, the plot adopts the gradient coloring scheme – the lighter the color is, the lower the fatality rate is.

Figures 2.8c1 and c2f provide very different perspectives on crash phenomena. They should be used to complement each other in drawing conclusions.

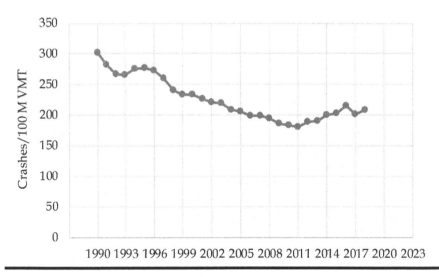

Figure 2.8b1 Motor vehicle crash rate per 100 million vehicle miles traveled. (Sources: Based on US DOT BTS National Transportation Statistics Table 2.17 at https://www.bts.gov/topics/national-transportation-statistics.)

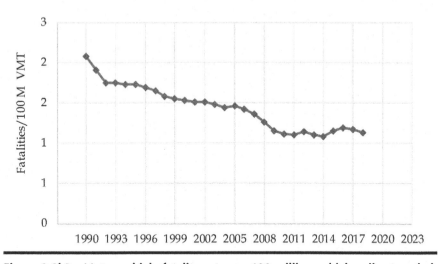

Figure 2.8b2 Motor vehicle fatality rate per 100 million vehicle miles traveled. (Sources: Based on US DOT BTS National Transportation Statistics Table 2.17 at https://www.bts.gov/topics/national-transportation-statistics.)

Figure 2.8c3 plots out geospatially the 2017 motor vehicle crash that occurred in the State of Maryland. Each dot represents a crash site. Such plots offer visually vivid location observation on where crashes occurred and how they are related to other geospatial reference points. From a magnitude standpoint, it is easy to identify what locations challenges are resided.

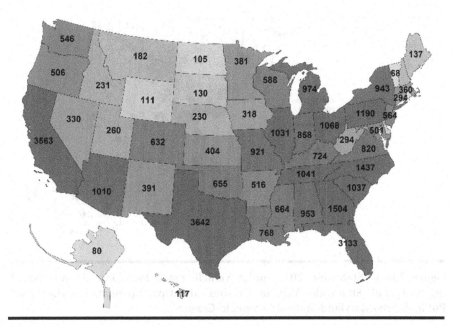

Figure 2.8c1   2019 Total motor vehicle fatality data. (Based on US DOT NHTSA Traffic Safety Facts Annual Report Tables – Table 125 at https://cdan.nhtsa.gov/tsftables/tsfar.htm#.)

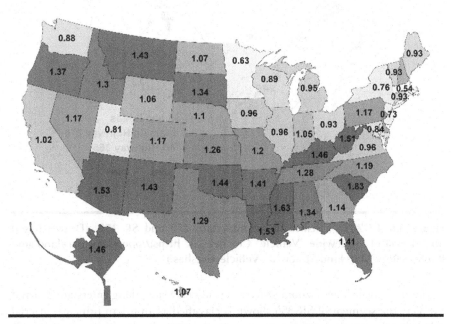

Figure 2.8c2   2019 Motor vehicle fatality per 100 million VMT. (Based on US DOT NHTSA Traffic Safety Facts Annual Report Tables – Table 125 at https://cdan.nhtsa.gov/tsftables/tsfar.htm#.)

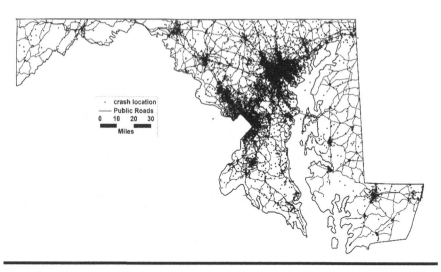

Figure 2.8c3 Statewide 2017 motor vehicle crash locations. (Drawn based on Maryland Statewide Vehicle Crashes at https://opendata.maryland.gov/ Public-Safety/Maryland-Statewide-Vehicle-Crashes.)

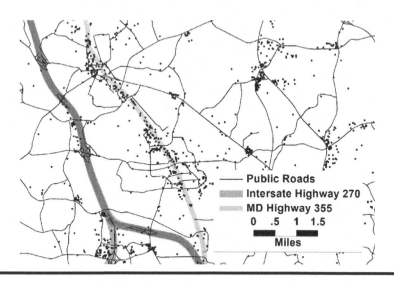

Figure 2.8c4 2017 Crashes along Interstate 270 and SR 355. (Drawn based on Maryland Statewide Vehicle Crashes at https://opendata.maryland.gov/ Public-Safety/Maryland-Statewide-Vehicle-Crashes.)

Figure 2.8c4 shows a zoomed-in review of crash sites along Interstate 270 and SR 355. The segment of SR 355 shown is classified as an urban principal arterial other than interstate and other freeways. It has at-grade intersections and significant numbers of access points with adjacent businesses. It is not surprising to see that SR 355 has denser dots as compared to the grade-separated free of access points I-270.

While visualization is appealing, quantitative analysis is always needed. They should supplement and complement each other

2.5.4.4 Integrated Safety Data Analysis

Crash causation data analysis as related to transportation infrastructure design and operations is complex. Often such analysis requires the integration of traffic data (e.g., flow rate and flow speed), roadway geometric (e.g., lane width, shoulder types, pavement condition), weather (e.g., precipitation), and behavior (e.g., distraction) data. Multivariate analysis is often deployed to tackle the issue. Additionally, significant research has been performed in crash prediction.

2.5.5 Planning Crash Data Analysis Example

To identify safety concerns for State Road 40 (SR40) from state line (MP 0.00) to Beaver Road interchange (MP 22.89), a 22.89-mile urban interstate.

Solution:
   Step 1: Data Collection and Basic Computation
      All cells in Table 2.2 with the shaded gray color background are data items that need to be collected.
      Cells without any shade (except title cells) are computed through formulas.
      a: Collect data in Table 2.2 covering cells b2 to e8.
      Centerline length, lane length, total VMT can be obtained from a State's roadway inventory database or from FHWA's Highway Performance Monitoring Information System (HPMS).
      Total crashes (cells d2 to d16) can be obtained from the Sates highway crash database.
      b: Collect data for cells: b10 to d16 and f10 to f16.
      c: Compute Total VMT for cells e10 and e16 by using the equation of b×f (centerline length×AADT) for the corresponding row.
      d: Cells c9, d9, and e9 are summations from their corresponding cells in rows from 10 to 16.
      e: Cell f9 (AADT) is obtained by dividing cell e9 (VMT) by b9 (centerline length). By the way, f2 to f8 are computed the same way originally.
   Step 2: Crash Rate Computation
      Cells from g2 to i16 are normalized crash data. Column "g"=d/b. It has the unit of crashes per centerline mile; Column "h"=d/c, having the unit of crashes per lane mile; and column "I"=d/e, providing the unit of crashes per 100 million VMT.

Table 2.2   SR 40 Safety Data Analysis Worksheet

	a	b	c	d	e	f	g	h	i
1	Coverage entity	Centerline length (Miles)	Lane length (Miles)	Total crashes	Total VMT	AADT	Crashes per thousand centerline mile	Crashes per thousand lane mile	Crashes per 100 million VMT
2	All roads in the state	221,409	673,161	15,612	5,739,734,955	25,924	71	23	272
3	Statewide urban road	60,735	194,352	12,400	5,260,926,435	86,621	204	64	236
4	Statewide rural road	160,674	478,809	3,212	478,808,520	2,980	20	7	671
5	Functional Class 1 – Urban	148	1,184	45	27,990,352	189,124	304	38	161
6	Functional Class 1 – Rural	125	750	39	13,834,750	110,678	312	52	282
7	Functional Class 2 – Urban	68	272	14	8,549,028	125,721	206	51	164
8	Functional Class 2 – Rural	104	416	23	9,314,968	89,567	221	55	247

(Continued)

Table 2.2 (Continued) SR 40 Safety Data Analysis Worksheet

	a	b	c	d	e	f	g	h	i
9	State Road 40 (Urban Interstate FC 1) (From MP 0.00 to MP 22.89)	22.89	137	11	4,393,695	191,948	481	80	250
10	Segment 1:MP 0.00 – 2.98	2.98	18	1	591,101	198,356	336	56	169
11	Segment 2:MP 2.98 – 4.92	1.94	12	0	447,511	230,676	0	0	0
12	Segment 3:MP 4.92 – 6.21	1.29	8	6	384,584	298,127	4,651	775	1,560
13	Segment 4:MP 6.21 – 10.45	4.24	25	2	846,647	199,681	472	79	236
14	Segment 5:MP 10.45 – 14.51	4.06	24	1	761,879	187,655	246	41	131
15	Segment 6:MP 14.51 – 16.88	2.37	14	0	402,343	169,765	0	0	0
16	Segment 7:MP 16.88 – 22.89	6.01	36	1	959,629	159,672	166	28	104

Step 3: Result Interpretation

The three sets of performance parameters in cells from g10 to i16 for the seven segments can be compared with statewide average conditions and other similar roads (e.g., All Urban Functional Class 1).

Among the seven segments, segments 3 and 4 have significantly higher crash rates than the other five segments. In addition, segment 3 crash rate is considerably higher than the statewide average, the statewide urban road average, and the statewide urban FC 1 average. The crash rate for segment 4 is lower than the statewide average, the same as the statewide urban average, and higher than the statewide urban FC 1 average.

Step 4: Conclusion

Based on what is revealed in Step 3, SR40 especially segments 3 and 4 should be planned for further evaluation and remediation for safety improvement.

2.5.6 *Highway Safety Improvement Program*

SAFETEA-LU established the Highway Safety Improvement Program (HSIP) as a core Federal-aid program. Since then, it has been continued with all subsequent surface transportation legislation. Each State Transportation Department leads the development of its statewide safety program called the SHSP. The SHSP identifies roadway features that consider hazards to road users and develops highway safety improvement projects based on crash data. Through partnering and collaboration, MPOs in urban areas adopt the statewide SHSP safety projects into their TIPs.

2.5.7 *Traffic Records Coordinating Committee*

Per 23 USC 405(c), the FHWA through 23 CFR § 1200.22(b) mandates each state DOT to establish a Traffic Records Coordinating Committee (TRCC) to provide a statewide forum to facilitate safety data collection and analysis. State TRCC memberships include representatives of the state highway agency, law enforcement entities, health professionals, and criminal justice. The TRCC is governed by a board and often has its own technical staff.

At the Federal level, the U.S. DOT TRCC composes members from FHWA, NHTSA, FMCSA, and the Office of the Secretary with a governing board. Its mission is to ensure all U.S. DOT operating administrations offer coordinated and efficient supports to state and local agencies.

2.6 Travel Demand Analysis

Travel demand analysis is performed to determine roadway infrastructure needs, assess operational options, and compare the effectiveness of various proposed strategies before building any facilities on the ground. Travel demand analysis is also used to answer "what if" scenario questions as related to proposed plans and policies.

2.6.1 Concepts and Definitions

2.6.1.1 Transportation Corridor

A transportation or highway corridor refers to a narrow swath of land between

WHY TRANSPORTATION DEMAND MODELING (TDM)

TDM is efficient in identifying transportation demand and gaining insights with "what if" scenario analysis without the need of actual field construction or actual implementation.

two or more population centers where a wide range of transportation facilities or modes has the potential to be built. The width of a corridor is not fixed (e.g., 10 miles wide), and it should be commeasuring with the goal of the project. The key is that the width of a corridor should not limit a comprehensive evaluation of all potential transportation solutions (e.g., a new highway, widening an existing road, or railroad) (Figure 2.9a).

Within the shaded area (corridor), there are both railways and NHS highways. However, there are no direct interstate highways linking the two regions. The task is to develop new railways or roads to connect the two areas. The width of the shaded area is not constant, and all the boundaries are approximate.

2.6.1.2 Roadway Network: Link and Node

Streets and highways are linear elements. The connections of those linear elements form a network. The intersection formed by two different linear elements is called

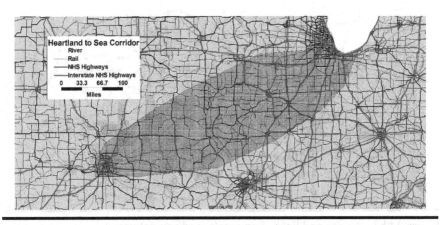

Figure 2.9a Transportation corridor illustration – seeking additional multimodal alternatives between the Heartland and the Great Lakes.

a network node. Network nodes include at-grade intersections, grade-separated interchanges, ramp connectors, and any other roadway character changing points. Geometric points include locations where the number of lanes changed (e.g., two-lane became four-lane). Operational environment change points may consist of posted speed change locations, median availability starting and ending points, presence of on-street parking, etc.

A node enables drivers to travel in a different direction and condition. A node offers an opportunity for drivers to change or adjust their travel behavior.

A roadway link refers to the segment of roadway between two nodes. A link typically possesses the same number of travel lanes, the same posted speed limit, and similar other roadway features (e.g., pavement type) (Figures 2.9b1 and b2).

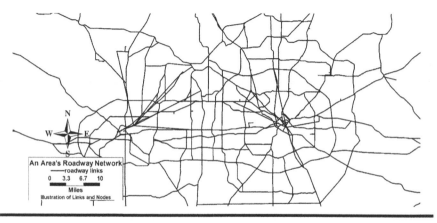

Figure 2.9b1   An area's roadway network illustration.

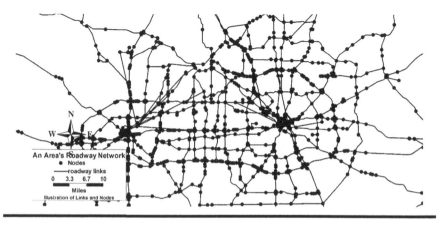

Figure 2.9b2   Nodes and links of an area's roadway network.

2.6.1.3 Roadway Link and Node Capacity

A roadway's ability to carry vehicles and people is one of its most fundamental functions. Roadway capacity is defined as the maximum number of vehicles or people that can sustainably travel through a uniform segment of the road (a link) within a specified time interval, under non-adverse weather or other environmental conditions, while conforming to posted speed conditions. Roadway capacity is expressed as the number of vehicles or people per travel lane, per hour, or per day traversing through a point.

For roadways classified as Interstate, Expressway, or Freeway, the capacity can be as high as 2,200 vehicles per lane per hour.

An intersection's capacity is defined as its ability to channel traffic to various paths such as through (continuing in the same direction), left turn, right turn, and U-turn. Each channelization path has a different capacity, which is controlled by both the geometric layout and traffic control (e.g., traffic signal).

Transportation Research Board's (TRB's) *Highway Capacity Manual* (HCM) is a practitioner's handbook for traffic analysis. The HCM is published and updated by the TRB of the National Academies of Science. The HCM contains theory, concepts, guidelines, and computational methods for assessing roadway capacity and services for all modes (vehicle, transit, bicycle, and pedestrian).

2.6.1.4 Traffic Flow Measurement and Characterization

The number of vehicles or people traversing through a point or a short segment of roadway in a given period is defined as the demand for the roadway.

Hourly volume is the total number of vehicles or people traversing through a point or segment in 1 hour.

Daily volume is the total number of vehicles or people traversing a point or roadway segment in one day.

2.6.1.4.1 Average Daily Traffic (ADT)

Average Daily Traffic is a daily average of less than one year's data. It is computed as the total number of vehicles or people counted divided by the number of counting days.

2.6.1.4.2 Annual Average Daily Traffic (AADT)

Annual Average Daily Traffic is the annual average daily number of vehicles or people using a highway. It is computed as the total number of vehicles counted in one year divided by the number of days in the year.

2.6.1.5 Travel Origin and Destination

A highway network offers paths for a traveler to travel from one place to another. The starting point of a trip is called origin, and the ending point is called a destination. The path of going from the origin to the destination is called a route.

2.6.1.6 Traffic Analysis Zones (TAZs)

A Traffic Analysis Zone (TAZ) is an enclosed geographical area marked by major roadways and geographical or political boundaries (e.g., rivers, city boundary). The size of a TAZ varies, depending on the overall project or program objective. Within a TAZ, the social and demographic characteristics are typically uniform. The uniformity character of a TAZ is often achieved by reducing the physical size of a TAZ. There are three categories of TAZs, as listed below.

2.6.1.6.1 Micro TAZs

An individual household or a few households (a neighborhood) are grouped together as a single uniform entity.

2.6.1.6.2 Meso TAZs

Households covering a larger geographic area (several neighborhoods) are grouped together as a single entity.

2.6.1.6.3 Macro TAZs

An enclosed area represents a large entity such as a city, a county, a region, a country, and so forth, depending on the overall modeling scale.

2.6.1.7 Centroid and Centroid Connector

A centroid is the center of a geometric area. For example, the centroid of a circle is the center of the circle. The centroid of a rectangle is the intersection of its two diagonals (Figure 2.9c).

An area's centroid is used to represent the area as if anything occurring within the area occurs right at the centroid. For travel analysis purposes, any trip starting within any location inside a TAZ is treated as started at the centroid of the TAZ. And a trip that ends anywhere in a TAZ is treated as that it ends at the centroid.

A centroid connector is a hypothetical roadway link connecting a centroid to its surrounding roads.

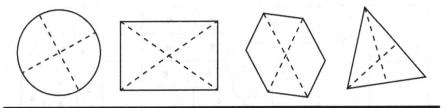

Figure 2.9c Illustration of centroids (intersections of dashed lines).

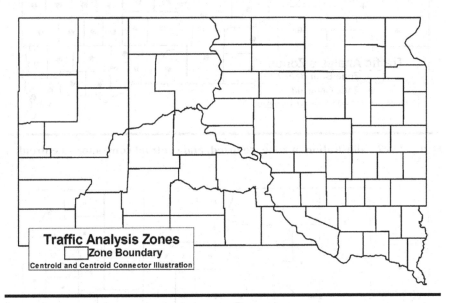

Traffic Analysis Zones
Zone Boundary
Centroid and Centroid Connector Illustration

Figure 2.9c1 Illustration of zone, centroid and centroid connector – zones.

Figures 2.9c1–c4 illustrate the relationships between TAZs, centroids, roadways, and centroid connectors.

Each geographical area defined by its boundary lines in all directions is called a TAZ.

Black dots are centroids of all the TAZs. The centroids act as trips' starting and ending points.

Figure 2.9c3 has the roadway overlaid on top of the TAZs and centroids. Some of the centroids are not located (connected) on top of a road. These non-connected centroids are not linked to any roadways.

To solve the non-connection issue between a centroid and a road, a "made-up" link (blue line) is created to connect a centroid to the adjacent roadway. These made-up links are called centroid connectors.

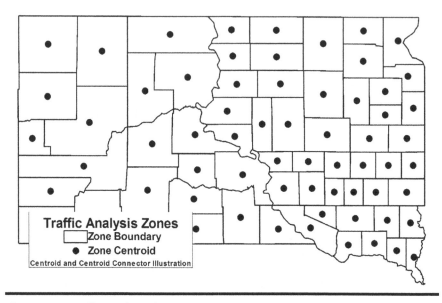

Figure 2.9c2 Illustration of zone, centroid, and centroid connector – centroid.

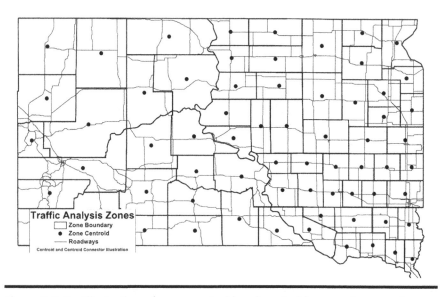

Figure 2.9c3 Illustration of zone, centroid and centroid connector – roadways.

2.6.2 *Classic Travel Demand Modeling*

Classic travel demand modeling is also known as four-step travel demand modeling. As its name implies, the procedure has four main steps.

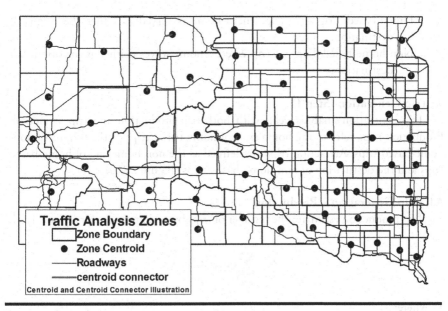

Figure 2.9c4   Illustration of zone, centroid, and centroid connector – centroid connectors.

Step 1: Trip Generation

Trip generation estimates the number of trips a TAZ generated. Trip generation covers two different trips – trip produced and trip attracted.

Step 2: Trip Distribution

Trip distribution estimates how trips generated in one TAZ are distributed to other TAZs. In other words, trip distribution connects trip origin and destination.

Step 3: Mode Choice

Mode choice reveals what mode (e.g., personal vehicle, public transportation, bicycle, walking, others) people are using or will use to travel.

Step 4: Trip Assignment

Trip assignment determines what route (which road or roads) a traveler will take to travel from one TAZ to another.

Results from the four-step modeling represent travel choices made by groups of homogeneous travelers through trip aggregation.

Several key terms used in travel demand modeling are listed below.

Trip – A trip refers to one-way travel from origin to destination.

Trip Purpose – The reason for making a trip is called trip purpose. Trip purpose can be work, shopping, visiting families, and others. The two most basic trip purposes are "work" and "non-work."

Home-based Trip – A trip where home is served as either the origin or destination.

Non-home-based Trip – A trip where a home serves neither as the origin nor destination.

A: Trip Generation (Step 1)

Trip generation analysis deals with both trip production and trip attraction. Production trips refer to the home ends of home-based trips or the origins of non-home-based trips. Attraction trips are defined as the non-home end of home-based trips and the destination of non-home-based trips (Figure 2.10a).

A1. Trip Generation – Production

Trip production estimates the number of trips by a trip purpose for a given TAZ.

$$P_i = \sum_{p=1}^{m} \left(R_{pas} \times H_{as} \right)$$

where:

P_i: number of trips produced covering trip purposes ranging from purpose $p = 1$ to purpose $p = m$ by travel zone i.

R_{pas}: trip rate for trip purpose p with auto ownership a and size of household s (# of persons in a household).

H_{as}: number of households with auto ownership a and size of household s (# of persons in a household).

Trip rate data $\left(R_{pas} \right)$ are typically obtained from household travel surveys. The FHWA's National Household Travel Survey (NHTS) provides such data for the entire nation. State DOTs and local MPOs also collect such data from local travel

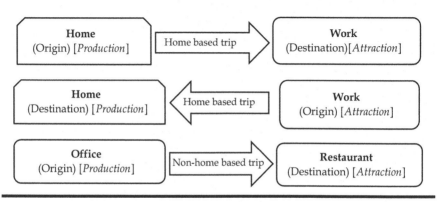

Figure 2.10a   Trip type illustration.

surveys. In the event where local data are not available, trip rate data can be borrowed from other localities for their initial usage.

Household data such as the number of households (H_{as}) with various social, economic, and demographic attributes are typically obtained from the U.S. Census Bureau and other state or local governmental statistical agencies.

A2. Example Computation

Task: Compute home-based nonwork person trips for TAZ 101, which is located in a nonurban area.

Step 1: Identify TAZ101 household data
A total of 345 households reside in TAZ101. These 345 households are further broken down into subgroups by auto ownership rates and household sizes, as listed in Table 2.3a. The household data are obtained from the local county population data center.

Step 2: Identify trip rate data for similar areas
The National Cooperative Highway Research Program Report 761 developed an average daily home-based nonwork person trip rates as listed in Table 2.3b. The basis of these trip rate data is the FHWA 2009 National Household Travel Survey.

Step 3: Compute home-based nonwork person trips generated by TAZ 101 by using $R_{pas} \times H_{as}$
Total number of trips can be computed by using the corresponding household data in Table 2.3a multiplying trip rate data in Table 2.3b as shown in Table 2.3c.

Step 4: Final result
Summarize all the data in Table 2.3c and obtain a total of 2,782.8 trips.
Zone 101 has produced a total of 2,782.8 $(P_{101}=2782.8)$ home-based nonwork trips.

Table 2.3a   Household Data for TAZ 101

	Household Size				
Autos	1	2	3	4	5 and 5+
0	12	6	12	0	0
1	10	19	14	14	0
2	5	34	21	32	24
3 and 3+	0	7	11	80	44
Total	27	66	58	126	68

Table 2.3b   Trip Rates for Non-Urban Area Home Based Non-work Person Trips

Vehicles	Household Size					
	1	*2*	*3*	*4*	*5+*	*Average*
0	1.2	3.3	5.1	8.1	10.3	2.6
1	1.9	3.6	6.7	9.5	10.3	3.5
2	2.0	3.6	6.7	9.5	12.1	5.6
3+	2.0	3.6	6.7	9.5	14.7	6.9
Average	1.8	3.6	6.7	9.5	12.9	5.1

Source: Courtesy NCHRP 761 Based on FHWA 2009 NHTS, www.trb.org.

Table 2.3c   Sample TAZ Trip Production Estimation

Autos	Household Size				
	1	*2*	*3*	*4*	*5 and 5+*
0	1.2×12	3.3×6	5.1×2	8.1×0	10.3×0
1	1.9×10	3.6×19	6.7×14	9.5×14	10.3×0
2	2×5	3.6×34	6.7×21	32	12.1×24
3 and 3+	2×0	3.5×7	6.7×11	9.5×80	14.7×44

A3. Trip Generation – Attraction

Attraction trips are defined as the non-home end of home-based trips and the destination of non-home-based trips.

The endpoints of non-home trips could range from an office building to a school, a business park, a cinema, etc. Common characteristics of non-home end zones (TAZs) include the number of people employed, square footage of various business facilities, and types of businesses (e.g., retail, wholesale, manufacturing). Table 2.3d lists several models used in estimating attraction trips. Similar to trip production computation, attraction trips are estimated by multiplying trip attraction rates with the base population or household data. Attraction trip production can be estimated through modeling equations.

A4. Example Computation

Task: Compute total number of home-based work attraction trips for Zone 101

Table 2.3d   Trip Attraction Rate (Person Trips Per Unit)

Number of MPO Models Summarized	School Enrollment		Employment[c]			
	Households[a]	Enrollment[b]	Basic	Retail[d]	Services[e]	Total
All Person Trips						
Home-based work						
16						1.2
Home-based nonwork						
2	1.2	1.4	0.2	8.1	1.5	
8	2.4	1.1		7.7	0.7	
2	0.7		0.7	8.4	3.5	
Non-home-based						
5	0.6		0.5	4.7	1.4	
	1.4			6.9	0.9	
Motorized Person Trips						
Home-based work						
8						1.2
Home-based nonwork						
1	0.4	1.1	0.6	4.4	2.5	
4	1.0		0.3	5.9	2.3	
Non-home-based						
6	0.6		0.7	2.6	1.0	

Source: Courtesy NCHRP 716, Table 4.4, www.trb.org.

a The number of households in a zone.
b The number of elementary, high school, or college/university students in a zone.
c Employment primarily in two-digit North American Industry Classification System (NAICS) codes 1–42 and 48–51 [Standard Industrial Classification (SIC) codes 1–51].
d Employment primarily in two-digit NAICS codes 44–45 (SIC codes 52–59).
e Employment primarily in two-digit NAICS codes 52–92 (SIC codes 60–97).

Step 1: Identify Zone 101 characteristics

	Household Size				
Autos	1	2	3	4	5 and 5+
0	12	6	12	0	0
1	10	19	14	14	0
2	5	34	21	32	24
3 and 3+	0	7	11	80	44
Total	27	66	58	126	68

A total of 345 households are within zone 101.

Step 2: Identify an attraction model or rate information.

Table 2.3d has a model equation in the form of "home-based work attraction trip = number of households × 1.2 trips/households."

Step 3: Compute the number of home-based work attraction trip by using the equation

Number of trips = 345 × 1.2 = 414 trips.

• Sources of Trip Rate and Frequency Data
Surveys (e.g., FHWA's NHTS)
• Population and Household Data
Census Bureau and Local Governmental Agencies (e.g., U.S. Census's ACS data)

A5. Balancing Production and Attraction Trips

Given that a trip always has a production end and an attraction end, total production trips should be equal to total attraction trips for a region. However, trip production and trip attraction estimation procedures are independent of each other. This leads to the reality that in practice, the estimated total trip produced and the estimated total trip attracted from all TAZs are generally not the same. This non-matching requires corrections to be made. The reconciliation of production and attraction trips is called trip balancing.

Trip balancing typically takes one of the three approaches listed below.

1. Holding the number of production trips constant

   With this approach, the total number of production trips is held constant, and the number of attraction trips is adjusted to match the number of production trips.

2. Holding the number of attraction trips constant

   With this approach, the total number of attraction trips is held constant, and the number of production trips is adjusted to match the number of attraction trips.

3. User-specified criteria

The user-specified criteria approach is the most widely used tactic where both production and attraction trips are adjusted to be the average of the original production and attraction trips.

During trip balancing, not every zone's production or attraction trip numbers need to be changed. A user can lock down certain zones and adjust the remaining ones only.

B: Trip Distribution Estimation (Step 2)

Once trip production and attraction estimates are complete, the next step in the modeling process is trip distribution. Trip distribution connects the production and attraction endpoints and establishes trip origin and destination pairs. Trip distribution predicts the spatial distribution pattern of trips as illustrated in Figure 2.10b.

Trip distribution modeling often follows the Newtonian gravitational principle. The Newtonian gravitational law concludes that every object attracts every other object with a force that is proportional to the product of the two masses and inversely proportional to the square of the distance separating them. Mathematically, this relationship can be expressed as:

$$F = g\frac{M_1 M_2}{R^2}$$

where:
M_1: object 1 mass.
M_2: object 2 mass.

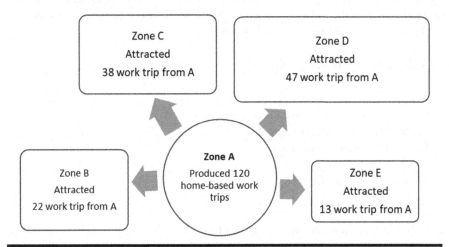

Figure 2.10b Trip distribution illustration.

F: force between M_1 and M_2.
R: distance between the center of M_1 and M_2.
g: gravity.

The application of the Newtonian gravitational principle to trip distribution is that the larger a pairing production and attraction TAZs is, the more trips between the pairing TAZs (larger OD flows). The shorter the distance between a paring production and attraction TAZs, the more trips occur between the pairing TAZs (larger OD flows).

Trip distribution following the Newtonian gravitational principle is mathematically described below.

$$T_{ij} = C_{ij} \frac{P_i \times A_j}{\sum_{j=1}^{m} \left(A_j \times C_{ij} \right)}$$

where:
T_{ij}: Number of trips produced in zone i and attracted to zone j.
C_{ij}: Travel cost factor between zone i and zone j.
P_i: Total number of trips produced from zone i.
A_j: Total number of trips attracted by zone j.
M: Total number of attraction zones.

The above model is called trip distribution gravity model.

Travel Cost Factor C_{ij}

There are a lot of approaches to derive the travel cost factor C_{ij}. However, regardless of what approach is taken, the derived travel cost factor must be calibrated. Model calibration is typically through trip length frequency distribution by trip purpose.

Below are two common methods to derive C_{ij}.

a. Gamma Function: $C_{ij} = kt^a e^{bt}$

where:
C_{ij}: travel cost factor between zone i and zone j.
T: travel time in minutes from zone i to zone j.
k, a, and b: model coefficients.

b. Power Function: $C_{ij} = t^a$

where:
C_{ij}: travel cost factor between zone i and zone j.
t: travel time in minutes from zone i to zone j.
a: model coefficient.

The result of a trip distribution analysis is the creation of a trip flow matrix (from-to), as illustrated in Table 2.3e.

C: Modal Choice (Step 3)

Modal choice refers to the mode a person takes to travel from his or her origin to his or her destination. People can walk, bicycle, drive, take public transportation, or car/vanpool, etc., to travel from one place to another.

Modal choice estimation can be as simple as using a standard fixed modal share lookup table to advanced complex models.

Given modal choices are always categorical and qualitative, typical models are discrete choice models (such as the multinomial logit and nested logit models).

The result of the modal choice analysis is the creation of a mode-specific person flow matrix (Table 2.3f).

Table 2.3e   Illustration of Total Home-Based Work Person Flow Matrix

From Zone I to Zone J	101	102	103	104	105	Total
101	1561.2	689.8	389.7	1896.4	798.2	5,335.3
102	982.1	826.1	218.3	2689.7	662.5	5,378.7
103	567.3	678.9	978.5	2895.7	128.3	5,248.7
103	120.6	168.9	1320.4	458.3	23.7	2,091.9
105	189.6	568.9	126.7	566.7	278.2	1,730.1
Total	3,420.8	2,932.6	3,033.6	8,506.8	1,890.9	19,784.7

Table 2.3f   Illustration of Home-Based Work Person Flow Matrix for the Bus Mode

From Zone I to Zone J	101	102	103	104	105	Total
101	0	0	128	120	29	277
102	0	0	0	0	12	12
103	120	0	0	224	18	362
104	24	0	220	0	12	256
105	26	13	17	19	0	75
Total	170	13	365	363	71	982

D: Route Choice (Step 4)

Route choice or trip assignment is the last step in the four-step modeling process. Route choice modeling identifies specific links and nodes a trip uses. Choices of links and nodes can be based on the time it takes to travel a link, the distance of a link, the cost of a link or node, the safety of link and node, and other factors.

Figure 2.10c1 illustrates links and nodes used to travel from location 1 to location 2 when the shortest distance specification is chosen as routing choice control. Figure 2.10c2 is based on the least travel time from location 1 to location 2. Depending on the controlling factors used in route choice modeling, links and nodes used to travel from the same origin and destination can be different.

Depending on its complexity, a route choice model may require data such as posted speed limit, roadway types, number of travel lanes, roadway capacity,

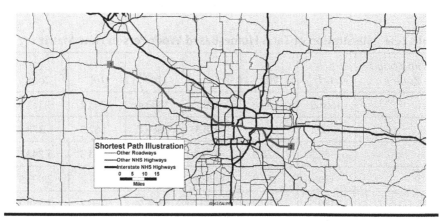

Figure 2.10c1   Route choice based on shortest distance criteria.

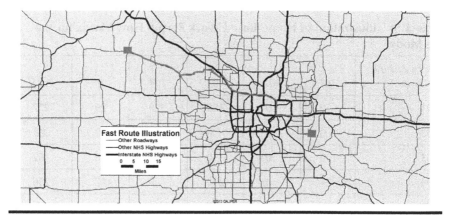

Figure 2.10c2   Route choice based on shortest travel time criteria.

intersection control, speed and volume relationships, transit schedules and fares, traffic signal systems, etc.

It is important to realize that link travel time depends on link traffic volume. When the traffic volume is high (congested), travel speed usually is low.

One of the most widely used route choice computation approaches is called the User Equilibrium Method (UEM). Under the UEM method, every user chooses a route that minimizes his or her travel cost (time, distance, etc.). No individual traveler can unilaterally change his or her route to further reduce his or her travel time once the system reaches its user equilibrium.

The UEM assumes each roadway system user has the perfect roadway network and travel information. However, this assumption is known not to be true.

The Stochastic User Equilibrium Method (SUEM) was developed to overcome the assumption that users have perfect knowledge of a roadway network. The SUEM assumes that no travelers have the perfect network information. An individual person's knowledge and his or her realization of network conditions vary. This new assumption ensures that for the same OD pair trips, there is a possibility that extremely low utility (e.g., long travel time) routes may still be taken by some travelers.

The last commonly used approach is called the System Optimum Solution (SOS). Under the SOS approach, the goal is to minimize total travel time over the entire roadway network by iteratively assigning flows to different links. At the equilibrium stage, it is possible for a given traveler to reduce his or her travel time by taking a different route, but the reduction is at the expense of other travelers on the network, and the overall travel time will increase.

E: Volume Speed Relationship

During traffic assignment, one of the often-used formulas linking traffic volume to speed is called the BPR (the Bureau of Public Roads) equation.

The standard BPR equation is:

$$S = \frac{S_f}{1 + a(v/c)^b}$$

where:
S: predicted mean speed
S_f: free flow speed
V: volume
C: capacity
a: 0.15
b: 4

However, when volume exceeds capacity, which is when a roadway operates in an oversaturated flow condition, the speed is no longer stable or sustainable.

The standard BPR equation does not function properly. Consequently, alternative formulas other than the original BPR equation have been developed to overcome this challenge.

2.6.3 Trip Generation, Distribution, and Routing

This section provides graphical illustrations of the traditional four-step modeling process (Figure 2.11a).

The size of the bubble in the chart represents the number of trips a TAZ (dashed line enclosed) generated and attracted. As shown by these split bubbles, the number of trips produced by a zone may be different from the number of trips attacked to a zone. Trip production and attraction data are generated with the first step in the four-step travel demand modeling process (Figure 2.11b).

The above desire lines show how trips generated in zone 106 are distributed to other zones. The thickness of each radiating desire line is proportional to the number of trips for each OD pair. The thicker the line is, the higher the number of trips from TAZ 106 to that zone.

These desire lines do not follow any roadways. At the present stage, routes (links and nodes) to get from zone 106 to other zones are not known. Trip distribution is step 2 of the four-step travel demand modeling process.

The above desire line represents vehicle trips only (modal choice). The modal choice analysis is step 3 of the four-step travel demand modeling procedure (Figure 2.11c).

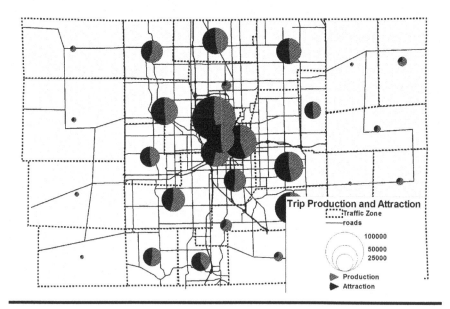

Figure 2.11a   Number of vehicle trips generated and attracted by TAZs.

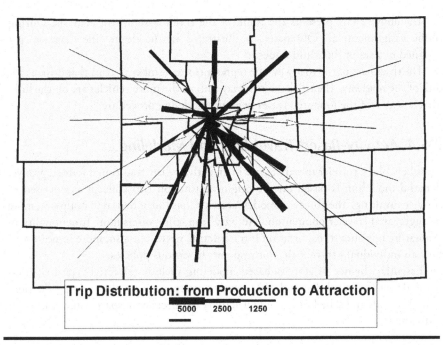

Trip Distribution: from Production to Attraction

5000   2500   1250

Figure 2.11b   Trip produced by Zone 166 and distributed to other zones.

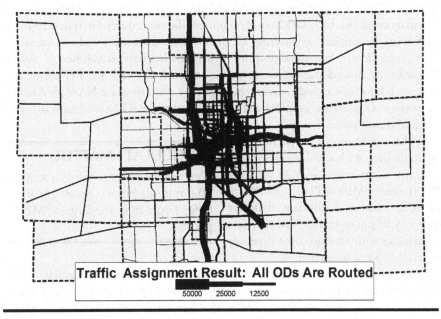

Traffic Assignment Result: All ODs Are Routed

50000   25000   12500

Figure 2.11c   Result of routing all OD trips on a roadway network.

The above chart presents the result of the traffic assignment analysis. During traffic assignment, all OD pairs, as illustrated in the desire line diagram, get assigned to a set of links and nodes.

The thickness of the roadway line represents the number of trips that utilize that link of the roadway. The thicker a roadway link is, the more vehicles are on the link. This is step 4 of the four-step travel demand modeling procedure.

2.6.4 Activity-Based Travel Demand Modeling

Travel enables a traveler to participate in an activity. The traditional four-step travel demand modeling is based on the origin-destination trip concept as opposed to activity thinking. The ability to model "activity" has the potential to evaluate a more disaggregated traveler population. "Activity" modeling offers more functionality in evaluating not only transportation infrastructure needs but also, more importantly, how an individual's travel behavior responds to various policies.

Essential themes of activity-based modeling include (a) activity participation is the driver for travel demand, (b) activity participation involves trip production, spatial spread, and scheduling, and (c) activity connects with space, time, and other individuals.

2.7 Transportation Conformity Analysis

The 1990 Clean Air Act Amendment legislation required the U.S. EPA, with the concurrence of the U.S. DOT, to develop specific procedures for transportation conformity determination covering air quality non-attainment and maintenance areas. Since then, a host of conformity regulations has been established to ensure that federally funded transportation activities, either through the FHWA or the FTA, do not impede a local governmental agency's effort to attain NAAQS. Among all six criteria pollutants, only PM and ground-level ozone (O_3) are relevant to highway and transit projects.

PM is a mixture of solid particles and liquid droplets with a wide range of sizes (effective diameters). NAAQS classifies PM as either PM10 or PM2.5. PM10 are particles with an equivalent diameter generally 10 μm or smaller. PM2.5 refers to particles with an equivalent diameter generally 2.5 μm or smaller.

PARTICULATE MATTER

The average human hair is about 70 μm in diameter – making it 30 times larger than the largest PM2.5 particle

Sources of PM include direct emissions from incomplete combustion, roads (dust, black smoke from a diesel truck, etc.), fields and open mines, factories, automobiles to wood-burning stoves. PM is also formed in the atmosphere through complex reactions of chemicals among sulfur dioxide and nitrogen oxides.

Ozone is a highly corrosive oxidant. Ozone at the ground level poses a serious health risk for all living organisms. Ground-level ozone is not emitted directly by industry operations or by internal combustion engines associated with automobiles. Instead, ground-level ozone is created by chemical reactions among volatile organic compounds (VOC) and nitrogen oxides (NO_x) in the presence of sunlight. The strong presence of sunlight during summer is one of the reasons why ground-level ozone is more a summer phenomenon than other seasons.

GROUND-LEVEL OZONE

- Not emitted by vehicles
- Produced through photochemical reactions in the presence of sunlight among VOCs and NO_x

VOCs are emitted from personal, commercial, and industrial activities and natural sources. With regard to motor vehicle operations, incomplete combustion byproducts contained in exhaust, gasoline vapors, and other gasoline leaks are the main contributors to VOC emissions. NO_x production is essentially all related to the burning of fossil fuel in industrial operations, utility production, and automobile usage.

Transportation conformity analysis is an effort to control vehicle-related emissions (PM, VOC, and NO_x) to achieve NAAQS standard.

2.7.1 State Implementation Plan

A SIP is a legal document developed by a state's environmental regulatory agency, outlining how a non-attainment or maintenance area plans to achieve NAAQS standard per Clean Air Act requirements.

A SIP includes emission estimates for baseline, episode, and future years. A baseline emission estimation is an estimate covering the current-year emission. An episode emission is an estimate of emission for the period where the actual field monitored ozone or PM exceeded NAAQS standard. Episode emission estimation is critical in establishing an emission budget – maximum emission allowed, for a region.

A SIP's goal is to reduce total emissions through different programs and strategies. State air quality agencies make decisions regarding amounts of emissions needed to be reduced from all sources.

With regard to mobile sources, limiting vehicle travel and reducing highway construction are potential means of reducing mobile source emission. However, doing so also affects the overall well-being of the economy, potentially increases congestion, and causes other harm to society. Often a State will choose to establish programs such as a Vehicle Inspection and Maintenance (I/M) initiative where registered vehicles are subject to periodic mandatory emission inspection. Those that fail to meet the vehicle emission standard are subject to repair. Promoting public transportation, carpooling, vanpooling in lieu of single-occupant vehicles, designating HOV, and enforcing HOV usage may also be deployed as SIP strategies.

Once a State's SIP is approved by the EPA for implementation, it becomes a legally binding agreement enforceable by Federal courts. However, a State can propose SIP revisions based on rules and regulations established by the EPA.

Figure 2.12a shows the cover page of the San Joaquin Valley Supplement to the 2016 State Strategy for the State Implementation Plan Revision adopted by the Californian Air Resource Board on October 25, 2018. This document outlines all

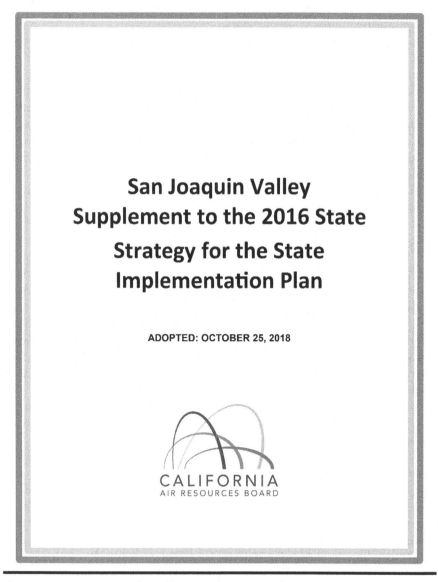

San Joaquin Valley Supplement to the 2016 State Strategy for the State Implementation Plan

ADOPTED: OCTOBER 25, 2018

CALIFORNIA
AIR RESOURCES BOARD

Figure 2.12a   San Joaquin Valley Supplement to the state SIP to attain ozone NAAQS.

the strategies, methods, and practices used to enable the San Joaquin Valley area to attain NAAQS standard for ozone.

Figure 2.12b shows the December 20, 2019 public notice with regard to the Texas Commission on Environmental Quality's effort to modify the 2015 adopted State SIP for the Boxer County 8 Hour Ozone Non-Attainment Area. While a

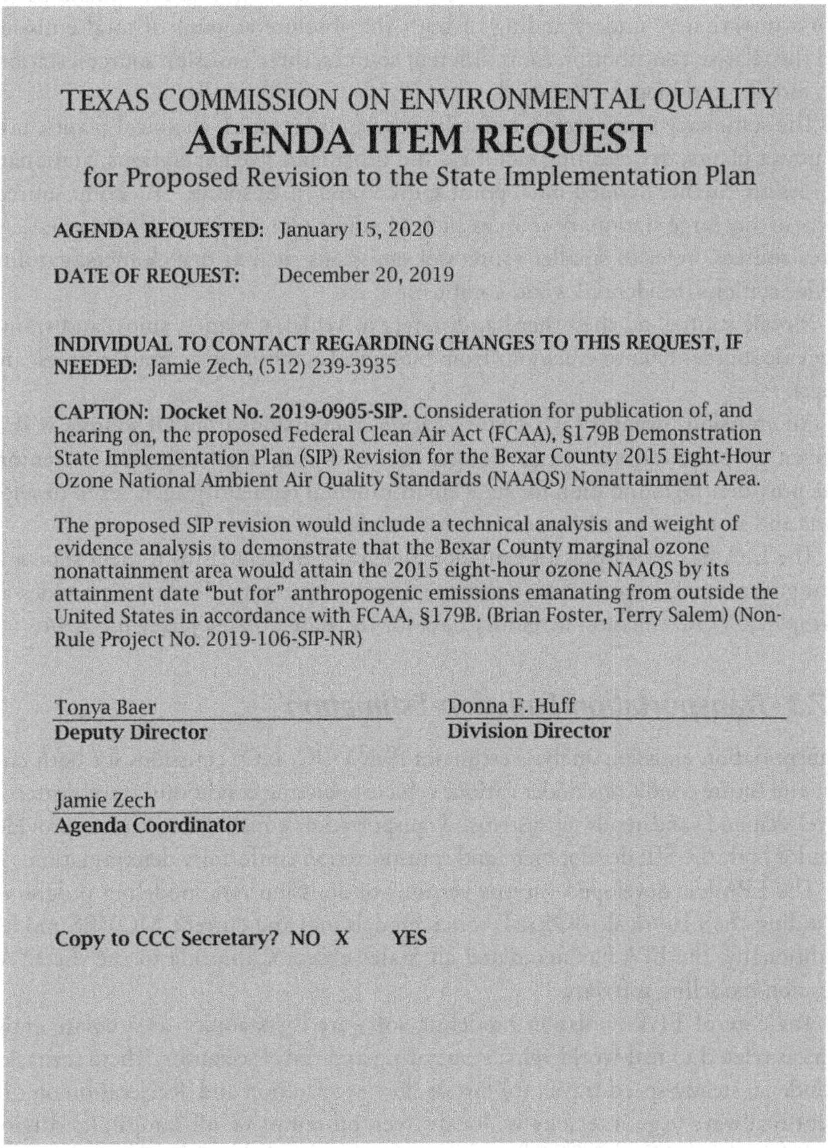

TEXAS COMMISSION ON ENVIRONMENTAL QUALITY
AGENDA ITEM REQUEST
for Proposed Revision to the State Implementation Plan

AGENDA REQUESTED: January 15, 2020

DATE OF REQUEST: December 20, 2019

INDIVIDUAL TO CONTACT REGARDING CHANGES TO THIS REQUEST, IF NEEDED: Jamie Zech, (512) 239-3935

CAPTION: Docket No. 2019-0905-SIP. Consideration for publication of, and hearing on, the proposed Federal Clean Air Act (FCAA), §179B Demonstration State Implementation Plan (SIP) Revision for the Bexar County 2015 Eight-Hour Ozone National Ambient Air Quality Standards (NAAQS) Nonattainment Area.

The proposed SIP revision would include a technical analysis and weight of evidence analysis to demonstrate that the Bexar County marginal ozone nonattainment area would attain the 2015 eight-hour ozone NAAQS by its attainment date "but for" anthropogenic emissions emanating from outside the United States in accordance with FCAA, §179B. (Brian Foster, Terry Salem) (Non-Rule Project No. 2019-106-SIP-NR)

Tonya Baer   Donna F. Huff
Deputy Director   **Division Director**

Jamie Zech
Agenda Coordinator

Copy to CCC Secretary? NO X   YES

Figure 2.12b   Public notice from TCEQ on SIP revision. (Sources: Courtesy of the Texas Commission on Environmental Quality, https://www.tceq.texas.gov/.)

State needs to follow what is prescribed in a SIP, the State can modify its SIP to make meeting the NAAQS more efficient by following appropriate legal steps and requirements.

2.7.2 Emission Inventory

A vital component of the SIP is its inventory of VOC, NO_x, and PM emissions. To gain a quantitative understanding of both the absolute amount of total emission and the relative contribution from different sources, three emission sources, stationary, mobile, and biogenic are used.

The stationary source refers to facilities such as factories, chemical plants, utility power plants, dry cleaning facilities, and other land fixed operations. Stationary sources are further divided into "point source" and "area source." A "Point source" refers to any large stationary sources such as power plants, chemical refineries, etc. "Area sources" refer to smaller sources of emissions such as dry cleaners, gasoline service stations, residential wood combustion, etc.

Mobile source, on the other hand, refers to vehicles, planes, ships, and trains. Biogenic sources refer to emissions from biological activities (e.g., grasses, trees, and crops).

An emission inventory is a quantification of emissions from all sources within a given geographical area (e.g., County, State) during a given period. Inventory data provides the foundation for state environmental regulatory agencies to develop plans and strategies to achieve air quality standards.

The EPA has developed an Emission Inventory System (EIS) to help state and local governmental agencies to perform emission inventory. Table 2.4 provides an example of VOC emission inventory data for the state of New Jersey for 2014.

2.7.3 Transportation Emission Estimation

Transportation emission analysis estimates PM, VOC, NO_x emissions for both current and future conditions under various vehicle operating conditions, travel patterns, travel demand, and roadway systems. Transportation emission estimation provides data for both the SIP development and transportation conformity determination.

The EPA has developed various versions of emission rate modeling programs, including the historical MOBILE series models and the current MOVES model. Additionally, the EPA has mandated all States except California to use the EPA's emission modeling software.

The core of EPA's emission modeling software is its ability to simulate emissions as related to real-world vehicle operating and travel scenarios. These scenarios include (a) steady-speed travel, (b) fast or slow acceleration and deceleration on different roadways (e.g., interstate vs. local street, on-ramps vs. off-ramps), (c) driving conditions (e.g., AC on/off), (d) weather condition (e.g., temperature, humidity), and (e) vehicle types.

Table 2.4 VOC Emission Inventory for New Jersey Covering Year 2014

Sources	Tons	Percentage
Fuel combined electric utility	285	0.1
Fuel combined industrial	413	0.1
Fuel combined other	7,470	2.7
Chemical & allied product manufacturing	202	0.1
Metals processing	70	0.0
Petroleum & related industries	136	0.0
Other industrial processes	3,810	1.4
Solvent utilization	60,638	21.5
Storage & transport	17,575	6.2
Waste disposal & recycling	2,777	1.0
Highway vehicles	32,845	11.7
Off-highway	31,527	11.2
Natural resources	102,877	36.5
Miscellaneous	21,198	7.5
Total	281,822	100.0

Source: Courtesy of EPA's 2014 National Emissions Inventory (NEI), https://www. epa.gov/air-emissions-inventories/2014-national-emissions-inventory-nei-data.

To facilitate the development of such modeling software, the EPA has conducted large-scale experiments that included tracing vehicles to gain a better understanding of how and when acceleration, deceleration, cruise control, and other driving patterns are carried out in the real world. By understanding such driving patterns, the EPA is able to simulate travel in a laboratory on a dynamometer.

Figure 2.13 shows EPA's "city test" cycle representing a city driving condition for light-duty vehicles. This test represents an average speed of 19.59 mph, with a total travel time of 1,369 seconds through 7.45 miles. The driving pattern shows a lot of stop-and-go conditions with frequent sudden acceleration and deceleration. When a cycle is driven on a dynamometer in a lab, emissions at various times can be measured and quantified.

Data from this type of testing enables the EPA to develop emission modeling software. The current mandated software for both emission inventory and

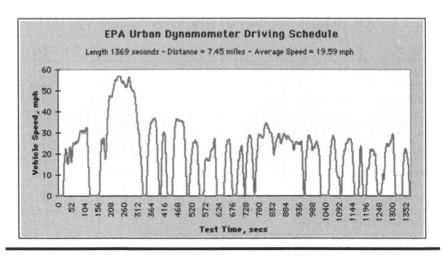

Figure 2.13 EPA city driving cycle for light-duty vehicles. (Courtesy of U.S. EPA Dynamometer Drive Schedules, www.epa.gov.)

transportation conformity determination is MOVES – Motor Vehicle Emission Simulator.

There are 13 vehicles types covered in the MOVES model. These vehicles are:

1. passenger cars,
2. passenger trucks,
3. motorcycles,
4. light commercial trucks,
5. intercity buses,
6. transit buses,
7. school buses,
8. refuse trucks,
9. single-unit short-haul trucks,
10. single-unit long-haul trucks,
11. motorhomes,
12. combination short-haul trucks, and
13. combination long-haul trucks.

There are four types of fuel covered in MOVES. And these are:

1. compressed natural gas,
2. diesel,
3. ethanol (E85),
4. gasoline, and
5. electricity.

Emitted pollutants modeled by MOVES include:

1. hydrocarbon (THC, NMHC, NMOG, TOG, VOC),
2. carbon monoxide (CO),
3. NO_x (NO, NO_2),
4. NH_3,
5. SO_2,
6. PM10,
7. PM2.5,
8. greenhouse gas (CO_2, CH_4, N_2O), and
9. air toxics (over 50 different species).

Emission sources from vehicles include:

1. vehicle running,
2. vehicle start,
3. extended vehicle idling,
4. fuel evaporation,
5. fuel permeation,
6. fuel vapor venting and liquid leaks,
7. vehicle refueling,
8. fuel vapor loss and spillage,
9. crankcase escape,
10. vehicle tire wear, and
11. vehicle brake wear.

Emissions estimated from the MOVES model can be quantified as either rate (e.g., mg per mile driven, mg per hour traveled) or total amounts of pollutants (e.g., lbs., tons). The rate for a given pollutant is grams per mile traveled, grams per hour of operation, or grams per engine start.

If the inventory function is chosen, MOVES generates total emission for motor vehicle activities within a given region. If this is the case, input to MOVES must also cover the complete vehicle travel activity vs. just travel profile.

If the emission rate mode is chosen, total emission can also be obtained by multiplying rates (g/mile, g/hour, g/start) with appropriate activity (e.g., miles, hours, starts).

$$Total_{emission\ (g)} = Rate_{g/mile} \times VMT_{total}$$

$$Total_{emission\ (g)} = Rate_{g/hour} \times VHT_{total}$$

$$Total_{emission\ (g)} = Rate_{g/starts} \times Starts_{total}$$

The rate method had always been the approach used by the MOBILE series modeling. The MOVES model has a new inventory function for inventory computation; thus knowing travel activity is critical.

In order for the MOVES model to output the appropriate emission rate or emission inventory data, appropriate MOVES input data must be developed. There are three major MOVES input data groups. The first one is fleet data – vehicle type, age, fuel, and usage. The second group of data is the vehicle travel pattern information by roadway types – VMT by roadway functional classes. The last group is climate data – minimum and maximum temperatures, humidity, and elevation.

2.7.4 Conformity Determination

The 1990 Clean Air Act Amendment stated that "No department, agency, or instrumentality of the Federal Government shall engage in, support in any way or provide financial assistance for, license or permit, or approve, any activity which does not conform to an implementation plan after it has been approved or promulgated ..."

The determination of whether an agency's transportation plan and program violate the above legislative intent is called transportation conformity determination. The FHWA and FTA make the conformity determination in consultation with the EPA. There has been a host of approaches for conformity analysis.

The first approach is known as the "build/no-build" test for the ozone issue. With the "build/no-build" test, an area must demonstrate that the "build" scenario will not produce more emissions than the "no-build" scenario for the analysis year.

A second test is known as the "less than 1990" test. An ozone non-attainment area must demonstrate that total emissions after implementing all planned transportation programs will be less than the 1990 base-year emission.

EMISSION MODELING

The more localized input data are, the more reflective of local emission rates or emissions are. Improving and localizing MOVES input data is one of the most critical tasks for mobile source emission estimation.

The latest method is the "budget test." With the "budget test" methodology, on-road mobile source emissions are compared with the "amount" (budget) of emissions allocated to the on-road mobile sources in a SIP. A transportation plan or program will only conform when emissions are less than or equal to the "budget."

2.7.5 Consequences of Not Meeting Conformity Determination

Failing to meet transportation conformity requirements is called conformity lapse. Conformity lapse has serious consequences for a region's transportation plan and program. During a lapse, the FHWA and FTA approval for new transportation projects will stop until revisions of either the SIP and/or the transportation plan (which enables a new conformity determination) are done. No new TIPs or projects can be approved for Federal funding or permitting during a lapse. Only safety, certain mass transit, air quality, and other projects such as highway noise abatement projects are exempt from the funding and permitting restrictions.

2.8 Financial Planning

Financial planning is a key component of transportation planning. Financial planning has three major components. The first is project cost estimation. The second is revenue projection. The last is project sequencing, which identifies when a project will start and complete.

2.8.1 Project Cost Estimate

Cost estimation at the planning stage is very difficult to develop because specific project data and information (e.g., concept, design, right of way) are lacking. Nevertheless, cost estimation is still needed. During planning, past project cost data for the region offer the best trending data and provide the basis for the current project cost estimations. The U.S. DOT's 23rd biennial Condition and Performance Report to Congress Exhibit A-1 outlines a national average cost for various highway projects (Table 2.5).

2.8.2 Revenue Estimates

The second component of financial planning is revenue estimation. Revenue estimation predicts types of funding, amount of funding, and when various funds will become available. Revenue sources include the Federal government, state government, local government, and private sectors. The types of funding can also be classified as highway user fees, taxes, and general revenue appropriation. Highway user fees and taxes typically include fuel tax, vehicle registration fee, heavy vehicle user fee, tire tax, licensed driver fee, and tolls. General revenue appropriation is fund appropriated from a general-purpose tax such as income tax, property tax, sales tax, etc.

Funding can also come from (a) issuing bonds, (b) private investment through public-private partnerships, (c) State-sponsored infrastructure bank to leverage borrowing, and (d) other mechanisms.

For revenue estimation, just as for project cost estimation, historical data offers benchmark information. In the absence of significant legislative changes or social and economic development, significant changes in revenue availability are unlikely.

Table 2.5 Typical Costs Per Lane Mile Assumed in HERS by Type of Improvement

Category	Reconstruct and Widen Lane	Reconstruct Existing Lane	Resurface and Widen Lane	Resurface Existing Lane	Improve Shoulder	Add Lane, Normal Cost	Add Lane, Equivalent High Cost	New Alignment, Normal	New Alignment, High
			Typical Costs (Thousands of 2014 Dollars per Lane Mile)						
Rural									
Interstate									
Flat	$1,993	$1,302	$1,128	$462	$86	$2,561	$3,551	$3,551	$3,551
Rolling	$2,234	$1,335	$1,298	$492	$142	$2,777	$4,493	$4,493	$4,493
Mountainous	$4,235	$2,924	$2,151	$728	$297	$8,646	$10,121	$10,121	$10,121
Other Principal Arterial									
Flat	$1,556	$1,042	$941	$371	$57	$2,052	$2,937	$2,937	$2,937
Rolling	$1,757	$1,071	$1,069	$413	$96	$2,197	$3,546	$3,546	$3,546
Mountainous	$3,412	$2,411	$2,072	$583	$126	$7,756	$8,931	$8,931	$8,931
Minor Arterial									
Flat	$1,423	$915	$877	$329	$54	$1,865	$2,618	$2,618	$2,618
Rolling	$1,718	$1,013	$1,091	$354	$99	$2,138	$3,372	$3,372	$3,372
Mountainous	$2,854	$1,871	$2,072	$486	$224	$6,547	$7,857	$7,857	$7,857

(Continued)

Table 2.5 (Continued) Typical Costs Per Lane Mile Assumed in HERS by Type of Improvement

Category	Reconstruct and Widen Lane	Reconstruct Existing Lane	Resurface and Widen Lane	Resurface Existing Lane	Improve Shoulder	Add Lane, Normal Cost	Add Lane, Equivalent High Cost	New Alignment, Normal	New Alignment, High
				Typical Costs (Thousands of 2014 Dollars per Lane Mile)					
Major Collector									
Flat	$1,499	$969	$905	$336	$69	$1,937	$2,617	$2,617	$2,617
Rolling	$1,640	$985	$1,018	$356	$93	$1,979	$3,220	$3,220	$3,220
Mountainous	$2,489	$1,541	$1,482	$486	$143	$4,191	$5,474	$5,474	$5,474
Urban									
Freeway/Expressway/Interstate									
Small urban	$3,356	$2,324	$2,645	$564	$103	$4,211	$13,784	$5,675	$19,373
Small urbanized	$3,608	$2,344	$2,736	$667	$137	$4,601	$15,117	$7,649	$26,114
Large urbanized	$5,754	$3,837	$4,238	$895	$517	$7,700	$25,826	$11,220	$38,303
Major urbanized	$11,509	$7,675	$8,224	$1,483	$1,034	$15,400	$64,219	$22,440	$85,845
Other Principal Arterial									
Small urban	$2,925	$1,974	$2,420	$473	$105	$3,579	$11,691	$4,474	$15,270
Small urbanized	$3,130	$1,998	$2,530	$559	$140	$3,878	$12,715	$5,520	$18,841
Large urbanized	$4,471	$2,929	$3,702	$703	$451	$5,675	$18,961	$7,577	$25,864

(Continued)

Table 2.5 (*Continued*)   Typical Costs Per Lane Mile Assumed in HERS by Type of Improvement

Category	Reconstruct and Widen Lane	Reconstruct Existing Lane	Resurface and Widen Lane	Resurface Existing Lane	Improve Shoulder	Add Lane, Normal Cost	Add Lane, Equivalent High Cost	New Alignment, Normal	New Alignment, High
	Typical Costs (Thousands of 2014 Dollars per Lane Mile)								
Major urbanized	$8,942	$5,857	$7,405	$1,135	$902	$11,350	$43,997	$15,154	$65,597
Minor Arterial/Collector									
Small urban	$2,155	$1,491	$1,831	$346	$76	$2,643	$8,562	$3,228	$11,019
Small urbanized	$2,258	$1,508	$1,848	$394	$93	$2,785	$9,050	$3,961	$13,520
Large urbanized	$3,040	$2,017	$2,527	$483	$253	$3,861	$12,820	$5,155	$17,594
Major urbanized	$6,080	$4,033	$3,822	$804	$507	$7,722	$43,997	$10,310	$54,445

Source: Courtesy FHWA Status of the Nation's Highways, Bridges, and Transit, Condition and Performance Report, 23rd Edition, www.fhwa.dot.ov.

2.8.3 Project Funding and Sequencing

The last component of financial planning is matching available financial resources with individual project needs. Different funding sources and funding types may have restrictions on their usages based on project types. For example, funding for Interstate Maintenance from the Federal government can't be used for non-interstate highway activities.

The matching of funding sources with various projects ensures that the transportation plan is realistic from a financial standpoint. Also, fund matching maximizes the utilization of all funds available to an agency.

2.9 Next Step

Projects contained in both the LRTP and the short-term TIP are developed by taking into consideration of all projects planned for a region. In other words, effects from all planned projects are considered holistically and systematically. The interdependencies of individual projects are studied and understood.

The delivery of the TIP is a milestone for an MPO. From there, state highway agencies will take over project activities and move forward with all remaining phases (e.g., engineering, NEPA, Design, Right of Way, Construction, Operations, and Maintenance).

The immediate step after Planning is the Class of Action (COA) determination. The COA determines whether any environmental impact analysis is needed for a project contained in the TIP and STIP. If a project needs to undergo any environmental impact analysis, the type of analysis required is determined.

2.10 Summary

Transportation planning is a critical step in the transportation project and program development process. An MPO is responsible for planning both its region's long-range plan and its near-term TIP. The Board of an MPO approves its long-range plan and its TIP.

A State DOT is responsible for developing its rural transportation plan and programs in cooperation with rural local elected officials. Additionally, the State DOT is required to develop statewide long-range plans, which may contain goals, strategies, and projects.

A State DOT is also required to develop a statewide TIP (STIP) by incorporating all its respective MPO TIPs and the rural TIP. After ensuring that the STIP complies with Federal laws and regulations, the State DOT is responsible for submitting its STIPs to the FHWA and FTA for a joint federal review and approval and keeping its MPOs informed with any Federal decisions and actions.

Both an MPO's TIP and a State's STIP need to be fiscally constrained, meaning projects contained in the TIP and STIP must have adequate funding. For NAAQS non-attainment or maintenance areas, an MPO's TIP must pass the transportation conformity determination meaning transportation projects for the region will not impede an area's ability to achieve air quality standards.

Transportation demand modeling provides an effective tool for assessing both present and future demand. It also offers the capacity for various scenario analyses with regard to effects from potential policy initiatives or project decisions before implementing such initiative or building an actual project in the field.

Once a state's STIP is approved by the FHWA/FTA, the state DOT and all the MPOs in the state can move projects in the STIP one by one into the next phase.

2.11 Discussion

1. What are the key deliverables of the transportation planning process?
2. Currently, the planning process focuses extensively on public involvement. Does this approach reduce the responsibility of the government and professionals involved concerning the final product? After all, medical doctors do not rely on the public to tell them what medicine to use for illness. Why does transportation rely so heavily on the public to tell professionals what to do? Also, does this consensus approach kill innovative and revolutionary ideas?

2.12 Exercises

1. Local governments own most public roads as measured by centerline length. True or False.
2. Interstates are free of access highways meaning all connections with interstate highways are through grade-separated interchanges. True or False.
3. There are three functional classes of highways based on functionality, according to the FHWA. These three classes are the arterial, collector, and local roads. True or False.
4. The National Highway System roads include all Interstate highways, most arterial, some collectors, and some local roads. True or False.
5. Highway functional classification balances mobility and accessibility. True or False.
6. A Transportation Management Area is designated by the U.S. DOT based on the geographical size of an urban area measured in square miles. True or False.
7. A Transportation Management Area is designated by the U.S. DOT based on an urban area's population. True or False

8. In the United States, the Department of Transportation determines urban area boundaries. True or False.

9. NAAQS stands for National Ambient Air Quality Standard. True or False.

10. Data used by the U.S. EPA to designate an area as air quality non-attainment are based on highly advanced air quality modeling software output. True or False.

11. Transportation planning regulation for air quality non-attainment areas is the same as regulations for air quality non-classifiable areas. True or False.

12. A state must have a State Implementation Plan if the state has an area designated by the EPA as air quality non-attainment. True or False.

13. A state's environmental regulatory agency is responsible for preparing the State Implementation Plan for air quality improvement purposes. True or False.

14. The creation of the Metropolitan Planning Organization is based on Federal legislation known as ISTEA. True or False.

15. An MPO operating in a TMA area must have a Board. True or False.

16. MPO board members are mainly elected local officials representing various local municipal entities. True or False.

17. Votes of MPO board members must be counted with the same weight. In other words, all votes are equal. True or False.

18. An MPO's Executive Director manages the MPO Board. True or False.

19. Most state DOTs are highway-centric. True or False.

20. The Federal Highway Administration is part of the U.S. Department of Transportation. True or False.

21. One of the most critical products an MPO produces is the Transportation Improvement Program (TIP). True or False.

22. A State may have many MPOs depending on the number of urbanized areas the State has. True or False.

23. A state DOT approves MPOs' TIPs. True or False.

24. A state DOT assembles all its MPOs' TIPs along with its state's rural TIP to establish a statewide STIP. True or False.

25. The FHWA and FTA jointly review and approve a State DOT submitted STIP. True or False.

26. Projects contained in an MPO's TIP do not need to have identifiable funding to support them because the TIP is a planning document. True or False.

27. Automobiles emit ozone, which may cause O3 NAAQS standard violation. True or False.

28. The Federal government, through the Federal Highway Administration, decides which highway should be built for each of the states in the United States. True or False.

29. A metropolitan planning organization has the authority to decide which roadway should be improved and when the improvement should occur. True or false.

30. An MPO's operation needs to follow the FHWA planning regulations and guidelines. True or False

31. Land-use planning and zoning is a Federal function. True or False.

32. Land-use planning and zoning is a local governmental function. True or False.

33. State transportation departments have the responsibility to carry out highway design and construction. True or False.

34. An MPO is subordinate to its State Department of Transportation from an administrative standpoint. True or False.

35. A State Implementation Plan (SIP) is developed by the State Department of Transportation to improve air quality. True or False.

36. Transportation conformity determination is made by the FHWA and FTA. However, the actual report preparation and analysis are performed by an MPO who is responsible for the area's TIP. True or False.

37. A road's capacity is the maximum number of vehicles that can pass through a point under prevailing weather and roadway conditions per hour. True or False.

38. A roadway demand is the actual number of vehicles traveling through a segment or passing a point of the highway within a given period. True or False.

39. A traffic data collection technician collected traffic count data for two full days (48 hours) with a total of 3,246 vehicle counts. The AADT can be computed as 3,246/2 = 1,623. True or False.

40. A traffic analysis zone (TAZ) can be as small as a couple of homes and as big as an entire city. Is the above statement correct? Yes or No.

41. A TAZ's centroid serves as the starting and ending point for all trips that occurred in the TAZ. True or False.

42. A centroid connector is an imaginary roadway link connecting a TAZ to a highway. True or False.

43. In travel demand modeling, the trip assignment step determines what roadway links and nods a trip will take. Methods to carry out the trip assignment can be very different. Under the so-called User Equilibrium Method, the assumption is that each road user has a perfect knowledge of the roadway travel condition and the roadway network. Is the above statement correct? Yes or No.

44. Vehicle emission modeling requires the usage of the EPA MOVES model to obtaining emission rates for all states except the State of California. True or False.

45. Transportation financing analysis includes the estimation of both transportation expenditure (where funds are spent) and revenue (where funds are from). True or False.

46. An MPO is responsible for its urban area's transportation planning. A State DOT is responsible for developing the Statewide Transportation Improvement Program (STIPs), and the Federal Highway Administration has the legal responsibility to review and approve the STIP. True or False.

47. Transportation conformity determination applies to all areas in the United States. True or False.
48. Compute the number of trips generated from the zone listed below by using trip rate data from Table 2.3b.

Autos	Household Size				
	1	*2*	*3*	*4*	*5 and 5+*
0	12	6	12	0	0
1	10	19	14	14	0
2	5	34	21	32	24
3 and 3+	0	7	11	80	44
Total	27	66	58	126	68

49. What are the most common data sources for trip rate information?
50. What are the most common data sources for social and demographic information?
51. Describe the four-step travel demand modeling in 200–250 words.
52. In travel demand modeling, the gravity model is often used for trip distribution modeling. What does the "gravity model" mean?
53. In travel demand modeling, the modal choice analysis is often through a multinomial logit model. What does "multinomial logit" mean?

Bibliography

2010 Census Urban Area FAQs, U.S. Census Bureau, https://www.census.gov/geo/reference/ua/uafaq.html.

Clean Air Act Overview, U.S. EPA, https://www.epa.gov/clean-air-act-overview/evolution-clean-air-act.

FHWA Functional Classification Guidelines, Concepts, Criteria and Procedures, FHWA Publication No. FHWA-HPL-13-026, 2013, Federal Highway Administration, Washington, DC, https://www.fhwa.dot.gov/planning/processes/statewide/related/highway_functional_classifications/fcauab.pdf.

FHWA Highway Statistics, HM10 and HM20, Federal Highway Administration, Washington, DC, https://www.fhwa.dot.gov/policyinformation/statistics.cfm.

Metropolitan Transportation Planning, 23 U.S.C. 134, GPO, https://www.govinfo.gov/app/details/USCODE-2011-title23/USCODE-2011-title23-chap1-sec134.

Nonattainment Areas for Criteria Pollutants (Green Book), U.S. EPA, https://www.epa.gov/green-book.

Patterns of Metropolitan and Micropolitan Population Change: 2000 to 2010, U.S. Census Bureau, https://www.census.gov/library/publications/2012/dec/c2010sr-01.html.

Transportation Conformity Determination, U.S. EPA, https://www.epa.gov/sites/production/files/2016-03/documents/58fr62188.pdf.

Transportation Management Area, 23 U.S.C. 134(k)(1)(A) and 49 U.S.C. 5303(k)(1)(A), GPO, https://www.govinfo.gov/content/pkg/USCODE-2017-title23/html/USCODE-2017-title23-chap1-sec134.htm and https://www.govinfo.gov/content/pkg/USCODE-2017-title49/html/USCODE-2017-title49-subtitleIII-chap53-sec5303.htm.

Chapter 3

Project Environmental Class of Action (COA) Determination

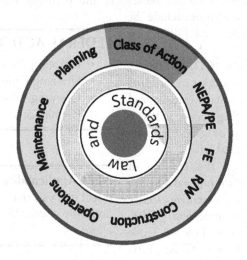

The key product of Transportation Planning is the Transportation Improvement Program (TIP), where a list of projects planned to be carried out for the next four years is generated. Before any project listed in a TIP can be further developed and implemented through design, right of way, construction, and other phases, the project must be in compliance with the National Environmental Policy Act (NEPA) in order for the project to be eligible for Federal funding and secure Federal environmental permits for construction.

3.1 National Environmental Policy Act

National Environmental Policy Act (NEPA) establishes a broad national framework for protecting the environment. The NEPA legislation was passed by Congress on December 23, 1969, and signed into law on January 1, 1970, by President Richard Nixon.

In its preamble, it states:

> To declare national policy which will encourage productive and enjoyable harmony between man and his environment; to promote efforts which will prevent or eliminate damage to the environment and biosphere and stimulate the health and welfare of man; to enrich the understanding of the ecological systems and natural resources important to the Nation; and to establish a Council on Environmental Quality.

NEPA requires that all branches and agencies of the Federal government give proper consideration to the environment before undertaking any major federal action that could significantly affect it. Federal actions could be granting permits, offering Federal financial assistance, and granting Federal permissions. Examples of Federal actions include the issuance of a Dredge and Fill permit (Section 404 permit) by the U.S. Army Corps of Engineers (COE), offering Federal financial assistance in building highways, airports, mass transit, and seaports, and granting access rights to interstate highways, or granting easements for using federal lands.

FEDERAL ACTIONS

- Granting Federal Permits
- Utilizing Federal Fund
- Granting Other Federal Permissions

NEPA authorized the creation of the Council on Environmental Quality (CEQ) within the Executive Office of the President to implement NEPA. The CEQ promulgated Title 40 Code of Federal Regulations (CFR) Parts 1500–1508 to carry out NEPA's intent.

NEPA Applicability – Environmental effect analysis due to Federal actions and only applicable to Federal agencies.

Title 40 CFR Parts 1500–1508 outlines three documentation classes to serve as administrative records of NEPA compliance. These three documentation classes are (a) Environmental Impact Statement (EIS)/Record of Decision (ROD), (b) Environmental Assessment (EA)/Finding of No Significant Impact (FONSI), and (c) Categorical Exclusion (CE). The environmental class of action (COA) determination identifies which one of the three documents is required for a project.

IF BAKER COUNTY WANTS TO CONSTRUCT A NEW HIGHWAY – WILL THE COUNTY NEED TO COMPLY WITH NEPA UNDER THE FOLLOWING CONDITIONS?

1. No federal funding is involved or will be involved.
2. No federal permits are needed on the environmental front.
3. The new county highway has its own independent utility and does not connect to the National Highway System.

The answer is no. This county highway project is not a Federal action. Consequently, NEPA does not apply.

3.2 Parties Involved in the NEPA Process

State and local government agencies are the most likely owners and initiators of transportation projects. Transportation projects contained in a STIP are typically owned by state Departments of Transporations (DOTs), meaning State DOTs are responsible for carrying all projects forward. Owners and initiators, like State DOTs, are called project sponsors.

For a highway project to move forward, it may need financial assistance from the Federal government. A bridge project may need a U.S. Coast Guard (Federal) bridge permit if the bridge is over a navigable U.S. waterway. A project impacting wetland will require the U.S. Army Corps of Engineers (COE) to issue a Section 404 permit. In cases like those, Federal action is sought, triggering the NEPA process.

Once the NEPA process is triggered, the relevant Federal agency becomes the lead agency in developing and approving the project. For highway and transit projects, the FHWA or the FTA becomes the lead Federal agency. A leading Federal agency is responsible for NEPA and all other environmental law and regulation compliance.

Project Sponsor – Most likely it is a state or local governmental agency. A project sponsor initiates the project.

Lead Federal Agency – the relevant Federal branch of government who is either granting permits or providing financial assistance to a project sponsor. Most likely the one that provides financial assistance will be the lead Federal agency. If there are multiple Federal agencies involved, other ones then become cooperating agencies.

Project sponsors (e.g., State DOTs) are responsible for carrying out the actual analysis and evaluation according to guidelines and procedures prescribed by the lead Federal agencies. The leading Federal agency reviews the analysis performed by a project sponsor, ensuring all processes and procedures are followed, and all analysis and documentation are in compliance with NEPA.

3.3 Class of Action Determination

Given that the NEPA is a Federal requirement, the authority for the NEPA COA determination rests with the lead Federal agency. For public highway projects, the Federal Highway Administration is the agency that has the authority to make the COA determination. For public transit projects, the Federal Transit Administration is the decision-maker.

3.4 Environmental Impact Statement

An EIS is reserved for a project where significant environmental impacts are expected from the proposed action. The CEQ defines "significant impact" in both "context" and "intensity" dimensions. Context refers to the natural, economic, social, geographic, physical settings of the action. Intensity relates to impact severity.

Purpose of the CE, EA, and EIS – Provide an administrative record documenting the undertaking is in compliance with NEPA.

An EIS details how a transportation project is developed, including the development of all reasonable engineering alternatives (no project is also an alternative) with environmental impact analysis of each alternative per relevant environmental laws and regulations.

Environmental impact analysis in an EIS describes the direct impact, indirect impact, cumulative impact, and mitigation commitments.

The direct impact is the actual immediate change to the environment, such as filling a wetland, tearing down a historical building, or clearing an endangered species habitat site.

Indirect impact refers to changes in social and environmental conditions that are not a direct result of an "action" but nevertheless are related to the action. For example, while the widening of a two-lane highway to four lanes linking a neighborhood to a major highway does not impact any critical habitat, the widening of the highway has enabled or expedited housing and neighborhood development in the area. In this example, the housing development is considered as an indirect impact by the roadway project.

Cumulative impact refers to the summation of impacts from all actions within a given region. Often each action itself has a minimal environmental effect. But when all these individual actions are added up, the total impact may become significant. For example, a single vehicle has virtually no impact on an area's air quality. However, if hundreds and thousands of vehicles are operating within the same area, the area's air quality may deteriorate.

Administrative steps in carrying out an EIS process are rigid. The sequence for the process is (a) issuing a Notice of Intent (NOI), (b) preparing and publishing a draft EIS for comment, (c) revising and publishing a final EIS, and (d) issuing a ROD.

The lead Federal agency publishes the NOI in Federal Register announcing the start of an EIS project. A draft EIS is prepared by the project owner, containing detailed information covering (a) needs, (b) purpose of the proposed action, (c) reasonable alternative solutions, and (d) both extent and severity of impacts to the environment for each alternative. Once a draft EIS is formally made available as announced in the Federal Register, the public and all other organizations and agencies can review and comment within a specified period. Upon closing of the public comment period, the sponsoring agency will address issues raised and prepare the final EIS.

The final EIS (FEIS) addresses comments received on the draft EIS. Most importantly, the final EIS identifies the "preferred alternative." Upon completion, the lead Federal agency announces the completion and availability of the FEIS via the Federal Register.

The ROD is the last step in the EIS process. Following the publication of the FEIS, the lead Federal agency issues the ROD. The ROD announces the selected alternative among all alternatives considered, states the rationale and basis for the selection, specifies the "environmentally preferable alternative," and offers complete information on the measures and actions necessary to avoid, minimize, and compensate for environmental impacts as binding commitments.

Keep in mind that NEPA does not mandate the selection of the "environmentally preferable alternative" as the overall "preferred alternative."

Transportation projects that require an EIS may include those listed below:

- New or extension of fixed guideway transit rail facilities
- New freeway or a new highway
- New bypass highways at new locations
- Major highway widening (e.g., two lanes to four lanes, four lanes to six lanes) projects through environmentally sensitive corridors.

3.5 Categorical Exclusions

Categorical Exclusion (CE) COA applies to Federal actions that neither individually nor cumulatively adversely affect the environment in a significant manner.

CE projects are categorically excluded from the requirements of preparing an EA or an EIS as an administrative record for NEPA compliance.

Leading Federal agencies can develop more detailed agreements with state governmental agencies about the further classification of CEs. Programmatic, Type I, and Type II CEs are often used to streamline the administrative record-keeping and the amount of evaluation and analysis performed.

Title 23 of the Code of Federal Regulation Section 771.117(c) and 771.117(d) outline a host of actions (types of projects) that can proceed without further NEPA approval by either the FHWA or the FTA. Some of these example activities include:

- utility installations,
- bicycle and pedestrian lanes, paths, and facilities construction,
- noise barrier construction,
- construction of fencing, signs, pavement markings, small passenger shelters, traffic signals, etc.,
- maintenance and improvements that are carried out within the existing right of way,
- installation of operating or maintenance equipment located within a transit facility and with no significant impacts on the site, and
- maintenance of a highway by resurfacing, restoration, rehabilitation, reconstruction, adding shoulders, or adding auxiliary lanes.

A lead Federal agency can elevate a CE COA determination when significant issues arise during the project development process.

3.6 Environmental Assessment

An EA COA determination is an action (a project) where the significance of a project's environmental impact is not known or understood at the time. A project under this COA designation requires the preparation of an EA to determine if and what further appropriate environmental documentation is required. The lead Federal agency may elevate the COA to an EIS based on the conclusion of the EA.

If an EA's analysis shows that there is no significant impact to the environment from the proposed action, a FONSI is prepared. A FONSI can be incorporated either as part of the EA or as a separate document to the EA.

COA Types – CEs, EIS, and EA

Who Makes the COA Determination – the lead Federal Agency. For highway projects, it is U.S. DOT's FHWA. For transit projects, the decision-maker is the U.S. DOT's FTA.

3.7 Engineering Alternatives for Environmental Analysis

Projects other than those classified as "programmatic" or "Type I CE" require the development of alternative solutions, including the no project alternative to accomplish the transportation goal. Alternatives are developed to enable environmental impact analysis with the goal of avoiding, minimizing, and mitigating impacts.

WHY ARE ENGINEERING ALTERNATIVES NEEDED FOR SOME CES, ALL EA, AND ALL EIS PREPARATION?

Engineering alternatives provide the footprint and layout for environmental analysis. For example, without knowing where a road will be, there will be no way to know how it will impact a wetland or neighborhood. Without knowing how wide (# of lanes) a roadway is, additional right of way can't be determined.

Environmental analysis for a given alternative is based on the proposed engineering solution. The environmental study provides feedback to the engineering design for further design enhancement and modification. The goal is that through this back and forth process, a project's ultimate transportation objective is achieved while environmental impacts are avoided, minimized, and mitigated by complying with NEPA.

3.8 Next Step

If a project is classified as Programmatic or Type I CE, the project can move forward without any further environmental analysis or evaluation. No further Federal NEPA approval is required.

For a project classified as Type II CE, EIS, or EA, the project will move to a phase called preliminary engineering and environmental impact evaluation (PE/EIE). During PE/EIE, engineering design alternatives, including a no project alternative, are proposed. These alternatives enable environmental impact evaluation. The goal of the PE/EIE is to gain approval from the lead Federal agency for the preferred design alternative. At the conclusion of the PE/EIE, engineering design for the preferred alternative is typically at 10%–30% completion.

3.9 Summary

The environmental COA determination is a pivotal decision-making point during the NEPA compliance process. The COA has a significant effect on a project's schedule, budget, and staff resource allocation. Any misclassification on the compliance requirement will lead to wasted resources and a prolonged schedule.

A project qualified as a Type I CE that is misclassified as a Type II CE will result in unnecessary resources being allocated to the project and will prolong the project schedule. If an EIS project gets classified as a Type II CE, it will lead to an unrealistic schedule for the project affecting budget and staff resource allocation in a significant way.

NEPA compliance is process-driven. Adherence to legally required steps is critical. Environmental advocacy groups and others often challenge NEPA decisions in Federal court. Past cases as related to such challenges have shown that courts often defer to agency experts on technical merits. Agency faults are often tied to not adhering strictly to the legally required administrative steps (e.g., 30-day public commenting period offered on the DEIS). Staff involved in the NEPA process must be well versed in the relevant federal legislation.

3.10 Discussion

1. What does a project's environmental class of action do? How does this determination affect a project's schedule?
2. Define the word "significantly" in the context of NEPA and environmental impact analysis. Explain it in 120 words.
3. A lead Federal agency makes the final decision on the COA determination for NEPA compliance purposes. If you are a state DOT project engineer, discuss how you would work with the lead Federal agency in the decision-making process.

3.11 Exercises

1. An Environmental Assessment (EA) Class of Action determination always leads to the preparation of a FONSI. True or False
2. The clear majority of highway projects are classified as Programmatic or CE. True or False
3. There are three basic environmental action classes. These are Programmatic/CE, EA, and EIS. True or False
4. An Environmental Impact Statement is typically just a few pages, as it is a statement. True or False
5. To carry out NEPA, "significant impact" analysis is covered with the dimensions of both context and intensity. True or False
6. Fill out the blanks for the following sentence.

    Environmental impact analysis under the context of NEPA addresses direct impact, _____ impact, and _____ _____ impact.

7. If the Federal government provides a grant to a village to build a new highway, this action is not a Federal action because the Federal government is not involved in the design and construction. True or False

8. There is no debate on how (methods) to assess the indirect environmental impact of EIS projects. True or False

9. NEPA is only applicable to Federal actions. True or False

10. For a bridge project over the navigable waterways of the United States, a Coast Guard bridge permit is needed. The bridge project will need to comply with NEPA regardless of whether Federal fund is used or not. True or False

11. The leading agency for a NEPA project is a state DOT given the state DOT is responsible for performing all the environmental analysis. True or False

12. A state DOT is planning to use Federal funds to develop a new 40 miles arterial roadway. The state's legislative body and the governor support the proposed roadway. As a matter of fact, the state legislature has passed specific legislation, and the governor signed it into law requiring the state DOT to build this new roadway with Federal fund. Because the project is authorized by State law, it does not need to go through the NEPA process anymore. True or False

13. An Environmental Assessment Class of Action determination is an action where the significance of environmental impacts from a project is not known or understood at the time. True or False

14. The EIS ROD identifies an "environmentally preferable alternative." The Final EIS identifies "preferable alternative." These two alternatives must be the same. True or False.

15. What is the legal basis for conducting environmental impact evaluation associated with Federal actions?

Bibliography

Categorical Exclusions, Council on Environmental Quality, https://ceq.doe.gov/nepa-practice/categorical-exclusions.html.

Determining NEPA Class of Action, FTA, https://www.transit.dot.gov/regulations-and-guidance/environmental-programs/determining-nepa-class-action.

The National Environmental Policy Act of 1969, https://www.energy.gov/nepa/downloads/national-environmental-policy-act-1969.

NEPA Compliance and Class of Actions, FHWA, https://www.fhwa.dot.gov/federal-aidessentials/companionresources/36nepacompliance.pdf.

Chapter 4

Preliminary Engineering

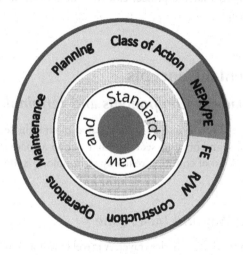

Engineers are challenged to design highways that meet the needs of different travelers under various conditions and constraints. Travelers encompass pedestrians, scooter and moped riders, cyclists, and motorized vehicle drivers. Performance metrics include travel time and distance, travel reliability, and travel safety. Constraints involve budget, environmental regulations, and public consensus.

Preliminary engineering (PE) design devises an engineering acceptable solution that avoids, minimizes, and mitigates unavoidable environmental impacts mandated by the National Environmental Policy Act (NEPA), has the public's support, and achieves the goal of the transportation project.

A project classified as either Programmatic or Type I Catrogical Exclusion (CE) in an approved Transportation Improving Program (TIP) can move forward without any further environmental analysis or evaluation.

A project classified as either Enviornemtnal Impact Statement (EIS), Environmetnal Assesment (EA), or Type II CE needs a PE design to support alternative analysis in compliance with NEPA and other relevant environmental laws and regulations.

It is only after the lead Federal agency (e.g., the Federal Highway Admnistration (FHWA), the Federal Transit Admnistration (FTA)) approves the CE, EIS, or EA/FONSI (Finding of No Signifcant Imapct) that the project can then be moved forward with its other phases and final design.

WHY PE AND NEPA ARE PERFORMED IN CONCERT

PE provides the layout and footprint where environmental analysis can be based on. In return, environmental analysis provides feedback to the PE where design can be modified to avoid and minimize environmental impact.

4.1 Fundamental Concepts

4.1.1 Basic Roadway Components and Terminologies

A roadway is a three-dimensional object consisted of straight segments and curved sections. It is these segments and sections that define a roadway's horizontal and vertical alignment. A list of common roadway terminology is illustrated and defined below.

4.1.1.1 Traveled Way and Travel Lane

Traveled way is a lane where vehicles travel. A travel lane is a delineated path traveled by a single profile of vehicles.

Traveled way includes lanes for both motorized and nonmotorized travel. The traveled way is most likely paved with either bituminous or Portland cement concrete.

4.1.1.2 Median

A roadway median is a strip of land situated between two traveled ways in the opposite direction. Median separates opposing vehicles – stopping errant vehicles from crossing-over and thus preventing head-on collisions.

4.1.1.3 Shoulder

A shoulder is a specially prepared surface area abutting from the edge of the traveled lane. The surface area to the right side of the rightmost travel lane is called the outside shoulder, and the area to the left side of the leftmost lane is called the inside shoulder. Shoulders can be either fully paved or partially paved.

4.1.1.4 *Rumble Strip*

Rumble strips are grooves in the otherwise flat pavement surface next to the outside of a travel lane that generate significant vibrations when tires run on top of them. These vibrations alert drivers to the fact that the vehicles they are operating just deviate from their supposed path. Corrective actions are needed. For example, a driver that has accidentally crossed the centerline will be alerted by the sound and vibrations a rumble strip generates. Types of rumble strips include transverse rumble strips, shoulder rumble strips, and centerline rumble strips.

4.1.1.5 *Curb*

A curb is a raised edge that delineates a traveled way. A curb can be vertical or sloped. Sloped curbs are more likely to be used in high-speed scenarios. When encountering an errant vehicle, a sloped curb normally doesn't cause major vehicle disturbance and has a minimal effect in redirecting an errant vehicle. On the other hand, a vertical curb typically causes a significant impact on an errant vehicle when the vehicle tire encounters the curb. Vertical curbs tend to redirect errant vehicles.

4.1.1.6 Curb and Gutter

A roadway curb combined with its adjacent pavement (flange) forms a gutter. Gutters act as stormwater conveyance channels. Working in concert with stormwater inlets, the curb and gutter system prevents roads from flooding. During construction, a curb and its adjacent pavement flange are often cast as a single piece with Portland cement.

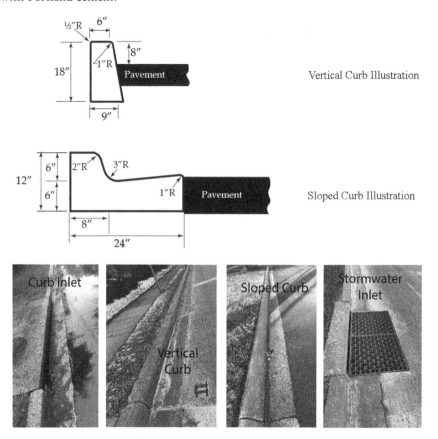

4.1.1.7 Roadway Ditch

A ditch is a channel dug within a road's right of way. It is typically parallel to the traveled way. Ditches are used for stormwater retention and conveyance. A ditch-type stormwater system is often used in rural areas.

4.1.1.8 Guardrail

A guardrail is a fence erected to prevent errant vehicles from crossing to the opposite side of a traveled way or from falling off a traveled way by redirecting vehicles back to their original traveled path. A guardrail system consists of two parts: a rail body and a pair of rail end.

4.1.1.9 Bicycle Lane

A paved lane situated next to a motorized travel lane intended for bicycle travel is called a bicycle lane. A bicycle lane can be designated as either dedicated or non-dedicated. A dedicated bicycle lane has the bicycle symbol or other visual indicators on the pavement while a non-dedicated bicycle lane lacks such pavement marking.

4.1.1.10 Sidewalk

Sidewalks are paved surfaces used by pedestrians that are spatially separated from motorized and bicycle lanes.

4.1.1.11 Lane Delineation Mark and Marker

Lanes are delineated with thermoplastic, paint, prism mirrors, and other physical objects either individually or in combination to guide drivers. While white lines indicate the same travel direction, yellow lines signify opposite directional travel on the other side. Drivers should not cross over any solid lines. Dashed lines signify that lane crossing is permitted.

4.1.1.12 Overhead Roadway Signs

Overhead Roadway Signs are signs displayed above-traveled lanes from either bridge span-type structures or cantilevered sign supports. Overhead signs are most often associated with high-speed roadways.

4.1.1.13 Roadside Signs

Roadside signs are displays installed on posts and placed outside the traveled way but within a roadway's right of way. Roadside signs are always placed away from the traveled way.

There are four basic types of roadway signs: guide, services/recreation, regulatory, and warning.

4.1.1.14 Variable Message Sign

Variable message signs are message display boards capable of presenting live messages to travelers. Contents of displayed messages are controlled remotely by Traffic Management Centers.

4.1.1.15 Driveway

A driveway is a short physical connector road between a roadway and its abutting property that enables safe entry to and exit from the property. Turning radius is a critical parameter in designing driveways.

4.1.1.16 Median Opening

A median opening is a gap in a roadway's median enabling vehicle crossing or turning.

4.1.1.17 Utilities

Utilities refer to infrastructure facilities providing non-highway travel-related services to the public but residing within a public road's right of way. Examples include buried water mains under the pavement and above-ground power transmission lines.

4.1.1.18 Crosswalk

A crosswalk is a striped narrow transverse path crossing a roadway designated for pedestrian use. Crosswalks are often associated with intersections. Occasionally, crosswalks may be stripped for middle block crossings.

4.1.1.19 Noise Abatement Wall

Noise abatement walls are structures erected by a highway agency to provide highway noise abatement. The location of a noise wall is typically parallel to the highway right of way line. Occasionally, it may be 2–3 feet inward (towards travel lane) from the right of way line, thereby providing access space for maintenance crews for repair and maintenance convenience.

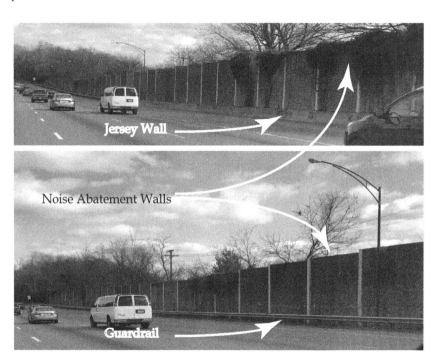

4.1.1.20 Landscaping Greeneries

Trees, bushes, and flowers that are planted to beautify highways are called landscaping greeneries. They are planted by highway agencies and communities.

4.1.1.21 Stormwater Retention and Detention Ponds

Stormwater retention and detention ponds are man-made water reservoirs with the sole purpose of storing stormwater runoff from a road. Detention ponds hold the water for a shorter period. It may be dry or wet throughout the year. Retention ponds always retain some stormwater throughout the year. Both detention and retention ponds are in relatively low-lying areas. They can be adjacent to the road or far away from the road depending on right of way cost and other factors. A pond always has an inlet and an outlet.

4.1.2 Station – Distance Measuring

A station is a unit of distance. One station is equal to 100 feet. Station numbers are used to reference locations of various roadway features in design plans and right of way maps. For example, where a curve starts and ends, where a drainage structure is to be constructed, and where a utility pipe is to be buried.

Stations expressed in a design plan take the form of Station A+B, where A represents the number of whole stations, and B represents a distance that is less than 100 feet. The starting point of a roadway is marked as Station 0+00, then every 100 feet forward, a new station is numbered sequentially.

For example, a 3″ diameter drainage pipe is to be buried from Station 116+56 to Station 202+15. The above station data not only show the starting points and endpoints of the pipe but also enable the calculation of the pipe length, which is 8,599 feet. The computation is done, as illustrated below:

$$((202 \times 100) + 15) - ((116 \times 100) + 56) = 8,599$$

Occasionally, a portion of a new route may be constructed along an existing route as illustrated in Figure 4.1. In this example, the new route, represented by the dashed line, starts from Station 0+0 to Station 4+96. This new route's second part is on an existing road starting from Station 11+50 to Station 26+50.

Station 4+96 and Station 11+50 represent the same physical location but are associated with two different linear reference systems (LRSs). In this case, the

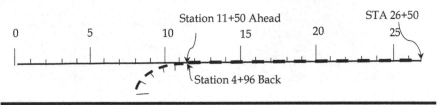

Figure 4.1   Illustration of stationing and station equations.

expression of "Station 4+96/Station 11+50" known as "Station Equation" is used. A station equation refers to an expression of two or more station numbers representing the same physical location but under different LRSs. For a given project, there may be many station equations involved. Attention must be paid to station equations ensuring proper distance computation. For the example illustrated in Figure 4.1, the total length of the new route can be computed as below:

$$\big((4\ \times100)+96\big)-(0+0)+\big((26\times100)+50\big)-(11\times100+50)=1,996$$

4.1.3 Tangents, Horizontal Curves, and Vertical Curves

A contiguous roadway is a connection between straight lines called tangents and various horizontal and vertical curves.

4.1.3.1 Tangents

A straight section of a roadway illustrated in Figure 4.2a is called a tangent. Tangents may consist of level roadway segments and uphill and downhill sections. A tangent cannot be part of any horizontal curves.

4.1.3.2 Horizontal Curves

A horizontal curve is a segment of roadway connecting two intersecting tangent roadway segments (intersecting tangents), enabling a smooth horizontal directional transition through a gradual turn. A horizontal curve is either circular, spiral, or circular-spiral combination curve (Figure 4.2b).

Circular curves are typically described by the following terms and characteristics, as illustrated in Figure 4.2ba1.

4.1.3.2.1 Basic Parameters

PI: The intersecting point (C) of two tangent roadways AC and BC is defined as the Point of Tangent Intersection (PI).

Figure 4.2a   An illustration of a roadway tangent section.

Figure 4.2b   An illustration of a horizontal curve.

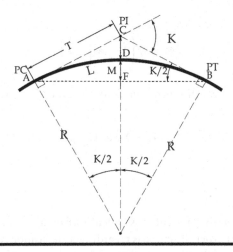

Figure 4.2ba1   An illustration of a circular horizontal curve.

PC: The starting point (A) of a curve with the tangent roadway AC is called Point of Curve (PC).

PT: The endpoint of a curve (B) with the tangent roadway BC is defined as the Point of Tangent (PT).

R: The radius is the distance from the center of the circle to the edge of the circle.

K: The central angle of the curve is K degree.

T: The distance from A to C or B to C is called tangent length.

4.1.3.2.2 Basic Relationships

Curve Length: Curve length (L) is the arc length from point A to point B and can be computed by the equation below.

$$L = \frac{KR\pi}{180}$$

where
L: arc length (feet) from A to B
R: radius (feet)
K: curve central angle (degrees)
External Distance: The length from C to D is called External Distance (*E*).

$$E = R\left(\sec\left(\frac{k}{2}\right) - 1 \right)$$

Middle Ordinate Distance: The distance from D to F is called Middle Ordinate Distance (*M*).

$$M = R\left(1 - \cos\left(\frac{k}{2}\right) \right)$$

Long Cord: The linear distance from A to B is called Long Chord (*LC*).

$$LC = 2R\sin\left(\frac{k}{2}\right)$$

4.1.3.2.3 Curvature

The sharpness of a horizontal curve is called horizontal curvature. The smaller a curve radius is, the sharper the curve is regarding its transition rate (Figure 4.2bb1). The measure of a curve's severity is defined by the angle subtended by 100-feet (30.5 m) arc length along the curve. The equation to calculate the angle for curve severity is:

$$D° = \frac{18,000}{R\pi}$$

where
$D°$: curvature expressed in degree
R: radius (feet)

4.1.3.2.4 Horizontal Curve Layout

There are five main horizontal curve layouts. The most commonly used layout is the single horizontal circular curve connecting two tangents, as illustrated in Figure 4.2bc1. This type of layout is commonly referred to as a simple horizontal curve layout.

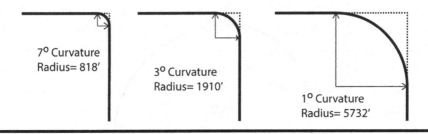

Figure 4.2bb1 **Curvature severity illustration.**

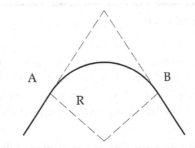

Figure 4.2bc1 **Simple circular curve.**

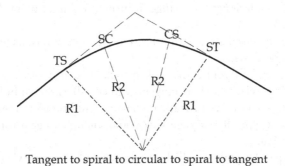

Tangent to spiral to circular to spiral to tangent

Figure 4.2bc2 **Spiral horizontal curve.**

The second horizontal curve layout type is the spiral curve illustrated in Figure 4.2bc2. The spiral curve layout offers a more natural and gradual transition from a tangent to a circular curve. Instead of the direct tangent to the circular curve, this layout is tangent to spiral, spiral to circular, circular to spiral, and spiral to the tangent (t-s-c-s-t). This added step not only further eases the transition from a tangent to a curve but also improves the aesthetic as the t-s-c-s-t layout is more eye-pleasing.

The usage of the spiral curve is originated from the railroad industry. The goal was to reduce track wear and tear by smoothing the tangent to circular curve transition. After its adoption for highway use, it was observed that during construction,

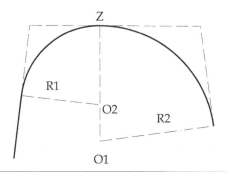

Figure 4.2bc3　Compound circular.

the layout of a spiral curve requires a substantial amount of extra survey work. Additionally, it requires more skilled labor for construction. Because of these extra demands, spiral curve usage in highways has been decreasing.

The third horizontal curve layout type is the compound circular curve depicted in Figure 4.2bc3.

A compound circular curve consists of two or more circular curves, with different radii, in succession connecting two tangents. A compound circular curve may deceive a driver on the curvature difference between the two curves due to the driver's inability to decipher the change in turning radius at point Z as illustrated in Figure 4.2bc3.

While the compound curve layout has a lot of applications, its usage should be limited to low-speed roads.

The fourth type of layout is the broken back circular curve, where two curves are connected by a relatively short tangent segment (d) as shown in Figure 4.2bc4. The broken back curve is aesthetically unpleasing. Additionally, a broken back circular curve layout typically does not have a long enough tangent runout for superelevation transition.

The last type of layout is called a reverse circular curve, where two successive circular curves in the opposite direction are used to connect two tangents

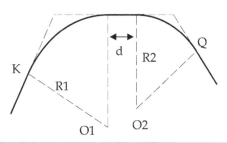

Figure 4.2bc4　Broken-back circular curve.

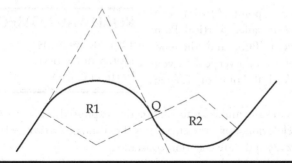

Figure 4.2bc5　Reverse circular curves.

(Figure 4.2bc5). The problem with the reverse circular curve is that there is no linear distance for any roadway cross profile transition (point Q in Figure 4.2bc5). As with compound circular curves, reverse circular curves should only be used in low-speed local roadways.

WHICH ONE IS A SHARPER CURVE? 2° OR 3° CURVE.

The 3° curve is sharper.

The 2° curve has a turn radius of 2,866 feet while the 3° curve has a turn radius of 1,910 feet.

4.1.3.3 Vertical Curves

A vertical curve connects two different grade roadway segments in situations such as hills and valleys as illustrated in Figure 4.2c1. Vertical curves always take a parabolic format guaranteeing a constant turning rate.

4.1.3.3.1 Basic Parameters

G_1 and G_2 are grades for uphill and downhill roadway segments. An uphill grade is always expressed as positive, while a downhill grade is expressed as negative.

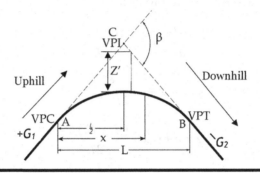

Figure 4.2c1　A vertical curve illustration.

1. The starting point (A) of a vertical curve is called Vertical Point of Curve (VPC), and the ending point (B) of a vertical curve is called Vertical Point of Tangent (VPT).

ROADWAY COMPONENTS

Tangent Sections
Horizontal Curves
Vertical Curves

2. Vertical curve formula: parabolic function expressed as $z = a + bx + cx^2$ where z is the elevation along the curve, x is the distance (horizontal) from point A (VPC), a, b, and c are equation constants.
3. Curve length (L) is the horizontal distance from VPC (A) to VPT (B).
4. The intersection (C) of two different grade tangent roadway segments is called Vertical Point of Intersection (VPI).

4.1.3.3.2 Basic Relationships

$$z = a + bx + cx^2$$

$$\frac{dz}{dx} = b + 2cx$$

$$\frac{d^2z}{dx^2} = 2c$$

The parabolic curve provides a constant rate of change ($2c$) on the roadway elevation transition, which accommodates both drivers' physical intuition and mental expectation.

1. Elevation along a vertical curve
   a. At the VPC (A): $x=0$, $z = a$
   b. $\frac{dz}{dx} = b + 2x$, when $x=0$, then $\frac{dz}{dx} = G_1 = b$
   c. Along the vertical curve anywhere:

$$\frac{d^2z}{dx^2} = 2c = \frac{G_2 - G_1}{L}$$

$$c = \frac{G_2 - G_1}{2L}$$

$$z = a + G_1x + \frac{(G_2 - G_1)x^2}{2L}$$

Elevation at any point along a vertical curve can be computed based on the above equation where "a" is the elevation of the VPC point.

2. Offset z' is the vertical distance from the original tangent elevation to the surface of the curve.

$$z' = \frac{|G_2 - G_1|\,x^2}{2L}$$

3. High and low points along a vertical curve (vertical clearance and drainage evaluation needs).

$$\frac{dz}{dx} = 0$$

$$x_{(min-max)} = L\left|\frac{G_1}{G_2 - G_1}\right|$$

4.1.3.3.3 Vertical Curve Types

Vertical curves can be grouped into the crest and sag vertical types. Within each type, there are three sub-situations, pending on both the direction and slope of the two connecting tangents, as illustrated in Figures 4.2c2 and c3.

4.1.4 Sight Distance

Sight distance refers to the maximum distance a driver can see in front of his or her vehicle. Stopping sight distance is the minimum distance needed for a driver traveling at the design speed to stop, avoiding striking an object on the roadway ahead.

Stopping sight distance has two components. The first part is the distance traveled by a vehicle in the time it takes for a driver to notice an object, realize the need

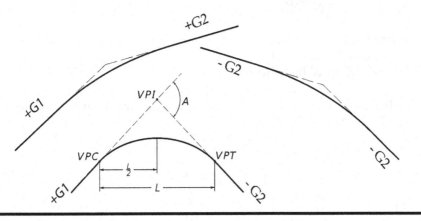

Figure 4.2c2   Crest vertical curve – three scenarios.

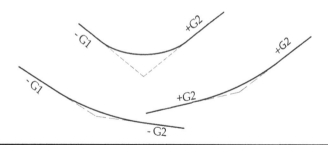

Figure 4.2c3 Sag vertical curve – three scenarios.

to stop, and apply the brake (reaction time). The second part is the distance traveled from when the brake is applied to the time the vehicle is stopped.

According to the American Association of State Highway and Transportation Officials (AASHTO) "A Policy on Geometric Design of Streets and Highways," 2.5-second reaction time, representing the 90th percentile of all driver reaction time, is adequate for use as a design parameter.

The following practical design equations are used.

a. Distance traveled during the first 2.5 seconds (representative reaction time) can be computed as:

$$d_r \left(\text{feet} \right) = V_{\text{speed}} T_{\text{time}}$$

$$= V_{\text{speed}} \left(\text{miles/hour} \right) \times \left(5,280 \ \text{feet/mile} \right)$$

$$\times 1 \left(\text{hour} \right) / 3,600 \ \text{seconds/hour} \times 2.5 \left(\text{seconds} \right)$$

$$= 1.467 V_{\text{speed}} \left(\frac{\text{feet}}{\text{second}} \right) \times 2.5 \left(\text{second} \right)$$

$$= 3.675 V_{\text{speed}}$$

where:
d_r: reaction distance (feet)
V_{speed}: design speed (miles/hour)

b. Braking distance for a vehicle traveling at design speed can be computed as:

$$V_2 = V_1 - a T_{\text{time}} = 0$$

$$T_{\text{time}} = \frac{V_1}{a}$$

$$d = V_1 T_{\text{time}} - \frac{1}{2} a T_{\text{time}}^2$$

$$d = V_1 T_{\text{time}} - \frac{1}{2} a T_{\text{time}}^2$$

$$d = \frac{V_1^2}{2a}$$

with unit conversations (miles to feet, hour to seconds), the above equation is converted to:

$$d_b = 1.075 \frac{V^2}{a}$$

where
d_b: braking distance (feet)
V: design speed (miles/hour)
a: deceleration (ft/s^2)

Total required stopping sight distance (d_t):

$$d_t = d_r + d_b = 3.675V + 1.075 \frac{V^2}{a}$$

For real-world practical design, a deceleration of 11.2 ft/s^2 is used. The 11.2 ft/s^2 represents a conservative estimate of most vehicles' braking capability for their tire under a wet pavement condition per AASHTO's design guideline.

4.1.5 Slope/Grade Effects on Stopping Sight Distance

Roadway slope/grade affects stopping sight distances. The additional gravitational force a vehicle experienced on a slope in addition to engine and brake force results in extra acceleration or deceleration.

4.1.5.1 Scenario 1: Uphill Travel

When a vehicle travels uphill, gravitational force slows the vehicle down. Stopping distance is reduced due to this extra deceleration force (gravitational force) in addition to the braking force (Figure 4.3a).
    Gravitational force is $F_{eg} = mg \, (\sin(\partial))$
    Extra acceleration along the slope: $-g \, (\sin(\partial))$
    Less distance to travel to stop due to gravitational force:

$$S_{ul} = \frac{3V^2}{2g \sin(\partial)}$$

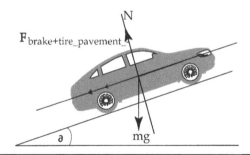

Figure 4.3a   Uphill gravitational force analysis.

where:

m: the mass of a vehicle

g: gravity ($32.17\,\text{ft/s}^2$)

α: road embankment slope

v: vehicle speed

S_{ul}: Less distance traveled for an uphill grade.

4.1.5.2 Scenario 2: Downhill Travel

When a vehicle travels downhill, the gravitational force acts as an extra acceleration force that the brake force needs to overcome (Figure 4.3b).

Stopping distance is now determined by:

$$S_{dm} = \frac{3mV^2}{2\left(F - mg\sin\left(\partial\right)\right)}$$

where:

m: the mass of a vehicle

g: gravity ($32.17\,\text{ft/s}^2$)

α: road embankment slope

v: vehicle speed

F: brake force

S_{dm}: distance traveled for a downhill grade.

DERIVING DOWNHILL TRAVEL STOPPING DISTANCE EQUATION

F is the brake force along the direction of slope

$$F - mg\sin\left(\alpha\right) = ma$$

$$a = \frac{F - mg\sin\left(\alpha\right)}{m}$$

$$V_{\text{stop}} = 0$$

$$V_{\text{stop}} = V - aT_{\text{stop}}$$

$$0 = V - \frac{F - mg\sin\left(\alpha\right)}{m}T_{\text{stop}}$$

$$T_{\text{stop}} = \frac{mv}{F - mg\sin\left(\alpha\right)}$$

$$S_{dm} = VT_{\text{stop}} + \frac{1}{2}\,a\,T_{\text{stop}}^2$$

Substitute T_{stop} into the above equation

$$S_{dm} = \frac{3mv^2}{2\left(F - mg\sin\left(\alpha\right)\right)}$$

The term $mg\left(\sin\left(\alpha\right)\right)$ in the denominator reduces the value of the denominator from when α is zero $\left(\text{no grade} - \text{flat}\right)$. Therefore S_{dm} increases as a result of the reduction in the denominator.

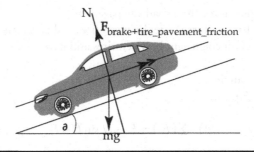

Figure 4.3b   Downhill gravitational force analysis.

4.2 Appropriate Horizontal and Vertical Curve Design

4.2.1 Design Appropriate Horizontal Curve

A vehicle traveling along the path of a circular curve experiences centripetal acceleration. Drivers sense this centrifugal force with a feeling that they are being pushed outward. This sensation causes drivers to steer more inwardly. Therefore, special attention must be given when designing horizontal curves, especially in moderate-to high-speed scenarios.

Several key interrelated factors, such as tire pavement transversal friction coefficient, superelevation, and vehicle travel speed, must be analyzed systematically to ensure safety.

Figure 4.4 illustrates various forces exerted to a turning vehicle. There are gravity force, pavement support force, centrifugal force, and the pavement tire transverse friction force involved.

where
F_c: centrifugal force
R: turning radius

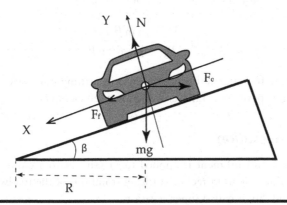

Figure 4.4   Force analysis for a turning vehicle.

N: normal support force from roadway pavement
F_f : transverse friction force (side) between tires and pavement $= N \times f$
f: transverse friction coefficient between tires and pavement (side friction)
m: vehicle mass
β: embankment angle
g: gravity (32.17 ft/s^2).

$$(1) \quad \Sigma(F_x) = F_c - N \sin \beta - F_c$$

$$(2) \quad \Sigma(F_y) = mg + F_f \sin \beta - N$$

$$(1a) \quad m\frac{V^2}{R} - N \sin(\beta) - Nf \cos(\beta) = 0$$

$$(2a) \quad mg + Nf \sin(\beta) - N \cos(\beta) = 0$$

$$(1b) \quad m\frac{V^2}{R} = N \sin(\beta) + Nf \cos(\beta)$$

$$(2b) \quad mg = -Nf \sin(\beta) + N \cos(\beta)$$

$$(c) \quad \frac{1b}{2b} = \frac{m\dfrac{V^2}{R}}{mg} = \frac{N \sin(\beta) + Nf \cos(\beta)}{N \cos(\beta) - Nf \sin(\beta)}$$

$$\text{simplifying (c)} \quad \frac{V^2}{Rg} = \frac{\sin(\beta) + f \cos(\beta)}{\cos(\beta) - f \sin(\beta)}$$

Divide both numerator and denominator of (c) by $\cos \beta$

$$(d) \quad \frac{V^2}{Rg} = \frac{\tan \beta + f}{1 - f \tan \beta}$$

Formula (d) is the fundamental equation linking turning radius (R), embankment angle (β), and tire-pavement transverse friction coefficient (f).

4.2.2 Superelevation

When a roadway embankment is applied to "a tangent to a curve to a tangent" transition situation, the unidirectional tilting (rotation) of the roadway cross slope (Figures 4.5a and 5b) is called superelevation.

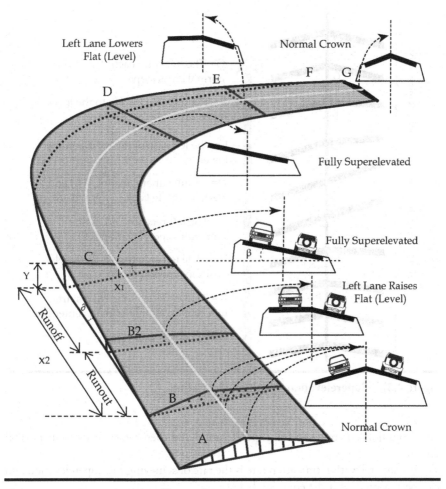

Figure 4.5a   Superelevation illustration.

Superelevation is measured and computed as:

$$e = \tan \beta = \frac{y}{x_1}$$

where
e: superelevation
y: embankment height
x_1: horizontal cross slope distance.

Superelevation transition is also known as cross-slope transition. Traveled way rotates along an axis of either the centerline of the median or the edge of the travel

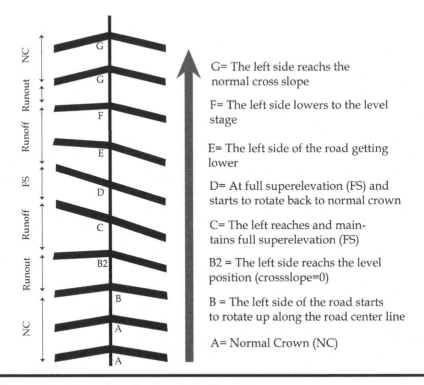

G= The left side reachs the normal cross slope

F= The left side lowers to the level stage

E= The left side of the road getting lower

D= At full superelevation (FS) and starts to rotate back to normal crown

C= The left reaches and maintains full superelevation (FS)

B2 = The left side reachs the level position (crossslope=0)

B = The left side of the road starts to rotate up along the road center line

A= Normal Crown (NC)

Figure 4.5b   Superelevation transition illustration.

lane from normal crown to full superelevation and then rotates back to its normal crown.

The superelevation transition rate is the rate of achieving full superelevation and can be analyzed with the equation below.

$$S_{tr} = \tan \partial = \frac{y}{x_2}$$

where

S_{tr}: superelevation transition rate (cross-slope transition rate for attaining full superelevation)

y: embankment height changed

x_2: horizontal distance traveled for the y amount of change.

Figure 4.5b shows that at location A, the traveled way is at its normal crown (normal cross slope). As it moves forward, the left traveled lane rotates in an upward direction along the centerline of the roadway (from B to C). By the location of B2, the left-side lane reaches a level position (runout). Upon reaching C (reached the runoff position), the entire left side of the traveled way has gained full superelevation. This full superelevation continues to D. From C to D, the full superelevation is maintained. From D forward, the rotation occurs in the opposite direction. By G, the lane rotates back to its normal crown.

The superelevation transition segment of the road has two parts (Figure 4.5b). The first part is called runout where it enables the rotating lane to rise from a normal crowned to a point where the lane has zero (flat) cross slope. The second part is called runoff through which the rotating lane is smoothly transformed from the zero-cross slope (flat) to its full superelevated cross slope. The full superelevated cross slope can be larger than the cross slope of a normal crown.

4.2.3 Design Appropriate Vertical Curve

For both crest and sag vertical curves, the curve length must meet the following specifications: minimum sight distance, an acceptable rate of grade change ensuring the physical comfort of both drivers and passengers, and acceptable aesthetic valued by the public.

A parabolic curve with an equal vertical axis centered at the VPI is commonly used for design. Curve length meeting the minimum stopping sight distance typically meets all other needs.

4.2.3.1 Crest Vertical Curve

Per AASHTO's "A Policy on Geometric Design of Streets and Highways" guideline, the length of an appropriate crest vertical curve can be calculated based on the following two equations under two different scenarios.

4.2.3.1.1 Length of an Appropriate Crest Vertical Curve

Scenario 1

The crest curve length is greater than or equal to the stopping sight distance

$$L = \frac{AS^2}{100\left(\sqrt{2h_1} + \sqrt{2h_2}\right)^2}$$

where:
L: crest vertical curve length (feet)
A: algebraic difference between the two grades (e.g., +3% as 0.03, −2% as −0.02. The algebraic difference is $0.03 - (-0.02) = 0.05$)
S: stopping sight distance (feet)
h_1: height of a driver's eye (feet)
h_2: height of an object (feet) to be seen by the driver.

Scenario 2

The crest curve length is shorter than the stopping sight distance

$$L = 2S - \frac{200\left(\sqrt{h_1} + \sqrt{h_2}\right)^2}{A}$$

where

L: crest vertical curve length (feet)

A: algebraic difference between the two grades (e.g., +4% as 0.04, −3% as −0.03. The algebraic difference is $0.04 - (-0.03) = 0.07$)

S: stopping sight distance (feet)

h_1: height of a driver's eye (feet)

h_1: height of an object (feet) to be seen by the driver.

4.2.3.2 Sag Vertical Curve

In addition to all issues related to crest vertical curve design, sag vertical curves have several additional unique factors needed to be considered. The first is drainage needs. Given sag vertical curves are associated with low points along with a roadway profile, proper drainage design (water flows to a low point) to keep travel lanes free of stagnant water (e.g., rain, melting snow) must be considered for all sag curves. Additionally, a sag curve length should be able to accommodate a vehicle's headlight beam distance. Lastly, a sag curve's length should not be shorter than the stopping sight distance.

For the physical comfort of drivers and passengers, the length of a sag vertical curve needed is only about half of the sage curve length required under the headlight distance criteria. The headlight distance criteria control the sag vertical curve length needed.

Proper sag vertical curve lengths for various grades and roadway design speed can be calculated by using the formula listed below according to AASHTO's "A Policy on Geometric Design of Streets and Highways" guideline.

Scenario 1

When the stopping sight distance is less than the curve length:

$$L = \frac{AS^2}{2(2.0 + S()]}$$

where

L: minimum sag vertical curve length when L equals to or greater than (=>) S

A: algebraic difference between the two grades

S: stopping sight distance.

Scenario 2

When the stopping sight distance equals to or is greater than the curve length:

$$L = 2S - \frac{400 + 3.5\ S}{A}$$

where

L: minimum sag vertical curve length when $L => S$
A: algebraic difference between the two grades
S: stopping sight distance.

4.2.4 Practical Design Consideration

While fundamental mechanistic equations provide the basis for research and development, in practice, empirical observations are often used in addition to the theoretical basis to establish design standards and specifications.

4.2.4.1 Superelevation

Superelevation should be less than 0.12 (12%) (Figure 4.5a: $\tan \beta =< 0.12$). The 12% slope is to prevent a vehicle from sliding along a roadway's cross slope especially in icy conditions.

4.2.4.2 Curve Radius

Minimum curve radius relies on the distribution of superelevation (e_{max}) and maximum transverse side friction factor (f_{max}) used in the formula below.

$$R_{min} = \frac{V^2}{15(0.01\ e_{max} + f_{max})}$$

where:
R_{min}: minimum horizontal curve radius in feet
V: design speed in miles/hour (mph)
f_{max}: maximum side tire and pavement friction coefficient
e_{max}: maximum superelevation. Its value equals to $\tan (\beta)$.

For any small defection (less sharp) curve, the curve should be long enough to avoid the appearance of a kink. AASHTO's Guideline prescribes that curve length on main highways should be no less than 15 times of design speed $(L_{min}\ (\text{feet}) = 15 \times S_{mph})$.

On high-speed highways, the minimum curve length should be no shorter than 30 times of design speed $\left(L_{\min}\left(\text{feet}\right)=30\times S_{\text{mph}}\right)$.

4.2.4.3 Tire and Pavement Transverse Side Friction Coefficient

For a given pavement spot, tire and pavement transverse side friction coefficient $\left(f_{\max}\right)$ is vehicle speed dependent. According to numerous research studies presented by AASHTO in its "A Policy on Geometric Design of Highways and Streets," the $f_{\max}$ could vary from 0.17 at 20 mph, 0.14 at 50 mph, to 0.08 at 80 mph.

4.2.4.4 Superelevation Transition Rate

Superelevation transition refers to the gradual rotation of traveled lanes from a road's normal crown to the superelevated position.

The maximum superelevation transition rate results in the shortest transition length. The current practice uses 0.78%–0.35% $\left(\tan\left(\partial\right)\right)$ in Figure 4.5a) for design speeds ranging from 15 to 80 mph. The higher the speed, the longer the transition is required.

The placement of the transition is between the runout and runoff. While typically the entire runout is on the tangent, it is desired that 60%–90% of runoff is also on the tangent according to AASHTO's guideline.

UNDERSTANDING SIGHT DISTANCE MEASUREMENT

The height of a driver's eye and the size of an object to be seen by the driver are critical factors to be considered. As vehicle design changes, the height of a driver's eye changes because heights of both the vehicle seat and vehicle itself may change. The height of the object to be seen by a driver varies depending on the size of the object (e.g., a squirrel vs. a deer). Obviously, the larger an object is, the easier it is to be seen, and the bigger impact it may have to a vehicle if there is a collision. The selection of heights for an operator's eyes and an object is based on actual field observations.

Eye height for a driver operating a truck is much higher than that for a passenger vehicle. This extra height enables a truck driver to see further. This compensates the additional stopping sight distance typically needed for trucks.

4.2.4.5 Vertical Curve

From a drainage standpoint, a minimum of 0.3% longitudinal grade is needed to facilitate stormwater flow longitudinally along a highway.

The minimum length of a sag vertical curve is the product of the algebraic difference between the two grades and 100. For example, if a down-slope grade is

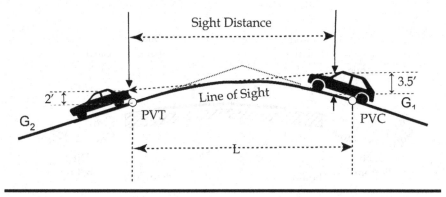

Figure 4.6 **Measuring sight distance.**

−3.5% and the uphill grade is +3.1%, then the algebraic difference between the two grades is $(-3.5) - (+3.1) = -6.6$. The minimum sage vertical curve length to link the two tangents is $100 \times 6.6 = 660$ feet.

4.2.5 Measuring Sight Distance on a Vertical Curve

For practical design purpose, the established standards are:

> Eyesight height: 3.5 feet
> Object Height: 2.0 feet

Figure 4.6 shows how a sight distance is measured for a crest vertical curve.

4.3 Roadway and Bridge Cross Section Design

A roadway cross section is the exposed surface when a road is cut transversely (right angles) to its travel direction by a plane. Features typically shown in a cross section are travel lanes and roadside elements.

Travel lanes refer to the number of lanes, lane width, and availability of bicycle lanes. Roadside elements include all other components within the right of way, such as shoulders, sidewalks, median, noise wall, drainage features, utilities, and guardrails.

4.3.1 Cross Section – Travel Lanes and Roadside Elements

Travel lanes refer to delineated paths for the traveling vehicles and bicyclists. Sidewalks are often physically separated from the vehicle and bicycle lanes, and they are considered part of the roadside components.

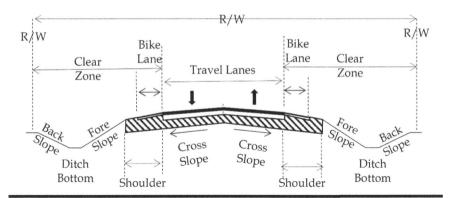

Figure 4.7a   A rural two-lane roadway cross-sectional illustration.

A single motorized vehicle travel lane is typically 12 feet wide. However, its width may be as narrow as 9.5 feet depending on traffic volume, the presence and percentage of trucks, and planned speed to be posted. Travel lane widths for dedicated bicycle lanes range from 4 to 10 feet. Sidewalk widths can vary from 4 to 6 feet.

A roadway shoulder extends the most outside travel lane further outward with reinforced soil, pavement, or both. It is a critical roadway element. A shoulder provides lateral support to the travel lane pavement, offers refuge area for broken down vehicles, and enables errant vehicles to recover. The width of a shoulder may range from 2 to 14 feet. A shoulder's cross slope is typically the same or slightly steeper than the adjacent lane.

Cross slope associated with traveled way enables stormwater to drain away from roadway surfaces and keep surfaces dry quickly (Figure 4.7a).

- This road is a two-lane rural road.
- The road has no median.
- This road has a paved bike lane in each direction.
- The road has a paved shoulder in each direction.
- The road handles its stormwater through a ditch and swale system.
- The space between the R/W line and the edge of the travel lane is called a clear zone (Figure 4.7b).
- This road is a four-lane rural divided highway with two travel lanes in each direction.
- The road has a depressed grassy median where its stormwater does not run to the traveled way.
- The road has paved outside shoulders on both sides of the traveled ways. However, inside shoulders are not paved. The inside shoulders are reinforced base material.
- The road handles its stormwater through a ditch and swale system on both sides.

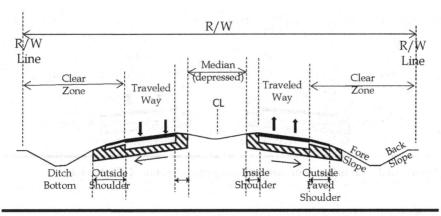

Figure 4.7b   A rural four-lane roadway cross section.

- Median stormwater is most likely routed to the roadside ditches by cross drain through underground pipes.
- The space between the R/W line and the outside travel lane edge is called the clear zone (Figure 4.7c).
- This road is a two-lane urban undivided roadway.
- It has a typical curb and gutter drainage design.
- The road has no median.
- The road has a bike lane in each direction.
- The road has a paved sidewalk in each direction.
- The road handles its stormwater through its curb and gutter system.
- The space between the R/W line and the lip of the gutter is called border zone (Figure 4.7d).
- This road is a four-lane urban curb design with two travel lanes in each direction.
- The road has a raised grassy median.
- The road has a bike lane in each direction.

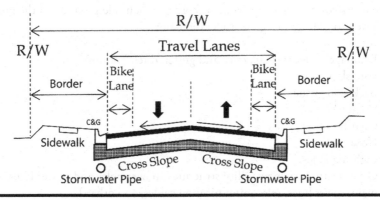

Figure 4.7c   An urban two-lane roadway cross-sectional illustration.

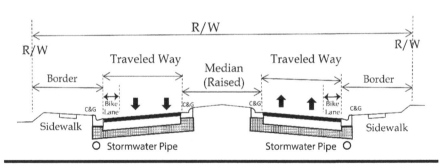

Figure 4.7d   An urban four-lane roadway cross-sectional illustration.

- The road has a paved sidewalk on each side.
- The road handles its stormwater through a curb and gutter system.
- The space between the R/W line and the lip of the gutter is called a border zone.

Roadside components refer to roadway medians and other elements residing within the space between the right of way (R/W) line and the outer edge of the travel lane.

The area separating the opposite traffic is called a median. The primary role of a median is to separate opposite directional traffic from head-on collisions, provide space for vehicle recovery, and serve as a refuge area for disabled vehicles or pedestrians trying to cross the road. A roadway's median can be raised, depressed, or flushed in reference to the adjacent shoulder. However, for effective drainage, most medians are depressed – where stormwater drains to the median vs. to the traveled way.

The area between the R/W line and the outer edge of the travel lane is known as the clear zone in a rural setting. The area between a curb lip in a curb and gutter setting and the R/W line is called the border zone.

A clear zone offers space for errant vehicles to maneuver and recover. No large physical obstructions such as trees, utility poles, and other rigid protruding objects should be present in a clear zone.

Elements covered in a clear zone or border zone (underground, on the ground, and above-ground) for design consideration are:

1. Drainage structure (e.g., curb and gutter, ditch bottom)
2. Fore-slope issue
3. Back-slope issue
4. Sidewalk
5. Guardrail
6. Noise wall
7. Lighting Poles
8. Other features such as utility structures (power poles and power lines, water main, sewer lines, telecommunication cables, and others).

The most important aspect of the roadside design is a) providing adequate space or a mechanism for errant vehicles to recover or be redirected back to the traveled way and b) offering stormwater conveying facilities.

Both curb gutter and ditch bottom drainage systems convene stormwater runoff offsite, thus keeping the traveled way dry.

For urban areas, given that right of way is typically expensive, curb and gutter drainage is used to reduce the right of way needs.

In rural areas, the land is often more abundant. Additional land acquisition is not an issue from a cost standpoint. Open ditch bottom drainage is used.

Depending on a roadway functional class and design speed chosen, specifications for clear zone width, shoulder widths, median widths, guardrail needs, and lane widths are different.

While AASHTO's "A Policy on Geometric Design of Highways and Streets" provides the minimum requirements for designing traveled way, AASHTO's "Roadside Design Guide" provides the minimum standards for designing roadside elements.

4.3.2 Bridge Basics

Bridges are an integral part of a roadway system. The functional part of a bridge where vehicles and pedestrians travel on is the bridge deck. A bridge deck is typically constructed by pouring concrete with steel reinforcement on top of beams. Bridge beams (e.g., steel girder, concrete box) are placed in the same direction as the traveled way on abutments, piers, and/or bents. Beams provide needed support to the deck. Abutments, piers, and bents support the beam.

An abutment is a supporting structure located at each end of a bridge. Highway bridge abutments also likely offer the earth retaining function in addition to supporting the beams.

For bridges that have more than one span, piers or bents provide the additional needed support to the beam between the two abutments. The distinction between a pier and a bent is its appearance. A pier typically has one column or shaft with one footing. A bent has two or more columns with each column supported by its own footing. There is no functional difference between a pier and a bent.

Piers and bents may be comprised of multiple piles. Piles can be constructed with precast piles driven into the ground by a pile driver. Piles can also be constructed with concrete poured into drilled shafts (boreholes drilled into the ground) to form in-situ piles in the field. In addition, piers can be formed from walls or columns resting on a spread footing. Piles and spread footing transfer bridge loads onto surrounding soil and bedrock to gain stability.

Figure 4.8a provides a bridge drawing illustrating basic bridge components and associated terminologies.

Bridge deck and beam are referred to as the bridge superstructure, while caps, columns, and piles are called substructure (Figure 4.8b).

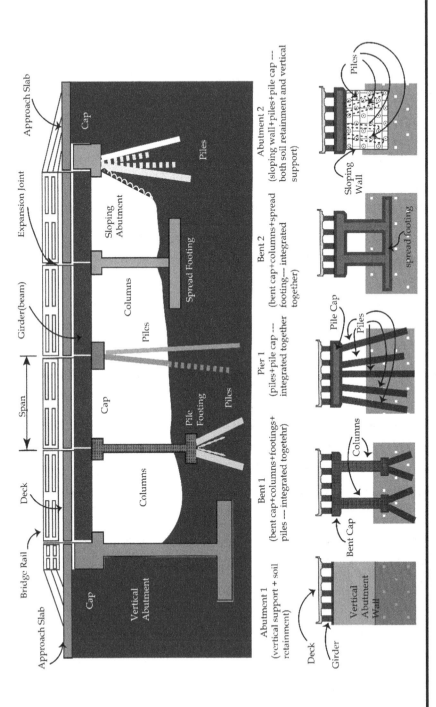

Figure 4.8a   A basic bridge drawing illustrating various components.

Figure 4.8b Pile driving vs. drilled shaft operations.

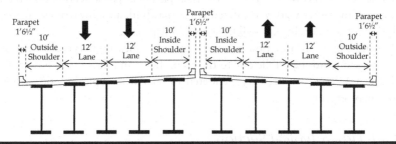

Figure 4.8c A two-lane per travel direction steel girder bridge cross section.

For any bridge over water, while its superstructure must meet the needs for vehicular and pedestrian travel, bridge pier design may also need to accommodate navigation needs. Both bridge horizontal clearance and vertical clearance (with both low and high tides if the waterway is tidal influenced) must meet the U.S. Coast Guard specification to prevent impacts and collisions with waterborne traffic (e.g., boats, ships, tankers ...).

As with any roadway design, a bridge is designed to ensure traveler safety. In addition to travel lanes, a bridge always has shoulders and bridge parapets (Figure 4.8c).

- This road has two separate bridges, with each bridge having two travel lanes.
- Each travel lane is 12 feet, and both inside and outside shoulders are 10 feet wide.
- The bridge is a steel girder bridge.

- The road handles its stormwater by draining all stormwater toward the outside parapet, where a valley is formed by the parapet and the roadway surface. Stormwater convening to a piping system either flows horizontally off the bridge or is drained down directly through scuppers.

Depending on a roadway functional classification, a bridge may also have physical barriers to protect pedestrians from errant vehicles.

4.3.3 Typical Sections

A typical section is a representation of roadway cross sections with similar or near-similar design features. A given roadway project may need more than one typical section to illustrate all the possible roadway cross-sectional configurations.

A typical section offers a glimpse of what a roadway looks like without knowing a roadway's complete plan, profile, and cross-sectional information. A typical section has four main parts. The first part shows the number of travel lanes in each direction, bike lane, sidewalk, and shoulder. The second is the median. The third component is its stormwater system. The final part of a typical section is its roadside features. If any other feature is offered (e.g., guardrail, a highway noise abatement wall, right of way fencing), it will be shown in this part.

During public information meetings or public hearings, typical sections are always presented to the public due to their simplicity and illustrative power.

Figures 4.9a–9c are drawings rendered as related to typical sections used for public involvement.

4.3.3.1 Urban Typical Section

For an urban area, two typical sections with various derivatives are typically used.

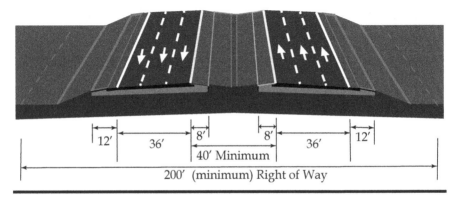

Figure 4.9a   An illustration of a proposed rural six-lane typical section.

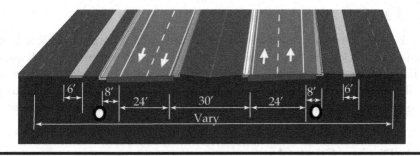

Figure 4.9b   An illustration of a proposed urban typical section.

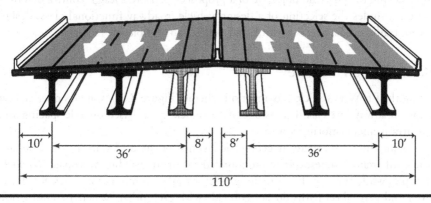

Figure 4.9c   An illustration of a proposed bridge typical section.

For high-speed design situations (design speed is 45 mph or higher), such as urban interstate, urban expressway and freeway, and other roadways, typical sections usually consist of:

- 12-foot lanes,
- a median that either contains a physical barrier or is wide enough to meet clear zone criteria,
- open ditch bottom or buried pipe stormwater system, and
- potentially roadside guardrails.
- Bicycle lanes and sidewalks are not included.

For low-speed roadway scenarios, local governmental agencies and the public often prefer the curb and gutter design over the open ditch system. The rationale is that the curb and gutter layout provides a clean and clear delineation for the road and is aesthetically appealing. Bicycle lanes and sidewalks are most likely part of the layout. Depending on the percentage of truck traffic, lane width may be narrower than the standard 12 feet.

4.3.3.2 Rural Typical Section

The biggest advantage of roadway design in rural areas is the availability of land. Consequently, clear zone and median spacing needs can be achieved through space, without the use of physical barriers or other protective measures. In addition, given the availability of land and sparsely populated housing units, drainage design is virtually all open ditch. Guardrails are often used to cope with hazardous natural conditions.

4.4 Design Control

Design control refers to the most fundamental design specifications that are critical in achieving the goal and objective of a proposed roadway. Design control determination includes the selection of a design vehicle, roadway functional class, design speed, traffic needs, access control, and pedestrian and bicycle facility.

4.4.1 Design Vehicle

Using the FHWA's vehicle classification scheme (Figure 4.10), vehicles operated on public roadways are classified as one of 13 vehicle types based on axle spacing and tire arrangement information.

Requirements on roadway geometric layouts for these 13 vehicles are different. These differences are especially pronounced for turning radius and space a vehicle occupies while turning. For a turning tractor trailer (combination truck 8–13), its front and rear wheel paths do not coincide with each other. When a combination truck turns, the circular arc performed by front of the tractor in addition to the off-tracking space of its rear wheel requires a wider lane than what otherwise is adequate on a tangent. This additional space is referred to as the "off-tracking" space.

For the design purpose, unless it is a local low-volume road where a passenger vehicle serving as the design vehicle is adequate, all other roadways should use the combination truck as the design vehicle.

For highways where there are significant recreational vehicles (RVs) towing boats and other trailers, the RV should be used as the design vehicle. RVs require wider spacing and a less sharp turning radius than combination trucks.

4.4.2 Roadway Functional Class

Typically, the functional class of a roadway is determined during the planning stage. The functional class of a proposed roadway is the most critical design control given its implication on design speed selection and access control.

Specifications on design criteria used by AASHTO's "A Policy on Geometric Design of Highways and Streets" are primarily based on roadway functional class. The guideline outlines specific criteria for local roads and streets, collector roads and streets, rural and urban arterial, and freeways.

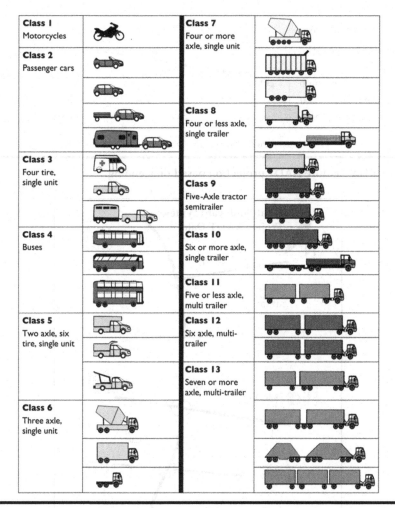

Figure 4.10   The FHWA 13 vehicle classification illustration. (Courtesy of the FHWA's Traffic Monitoring Guide, www.fhwa.dot.gov.)

State highway agency design guidelines also use the roadway functional class as their primary criterion.

4.4.3 Traffic

Planning's travel demand modeling produces AADT data for roadway links. For a project during its PE/EIE stage, a more focused traffic analysis through corridor/subregional traffic demand modeling, historical trending (Figure 4.11a), or other approaches is carried out. These additional analyses refine AADT and turn movement data for all interested roadway links and intersections.

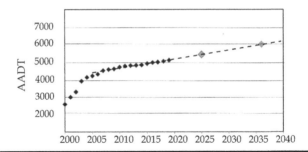

Figure 4.11a Future AADT projection based on trending.

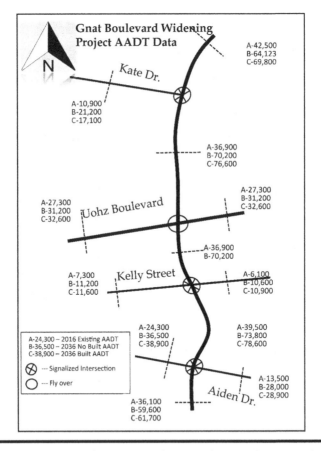

Figure 4.11b Illustration of AADT data for a roadway-widening project.

In these refinement analyses, link AADT and intersection/interchange turning movement data are estimated for (a) existing condition (current year), (b) no built (no project) conditions for opening and design years, and (c) built option (with the project) for both opening and design years.

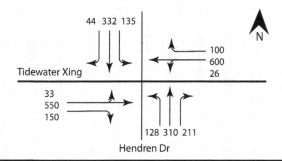

Figure 4.11c   Illustration of AM peak vehicle turning movements.

Figure 4.11b illustrates how AADT data are presented for various links. As shown, the project's route has five major links separated by intersections where each link has its own unique AADT.

For intersection movement, peak hour traffic movements on each link are estimated. Data on turning movement relies heavily on field observations. Figure 4.11c illustrates how intersection turning movement data are presented. For each link, movements are separated into through, left, and right turns.

4.4.3.1 Understanding AADT and Its Relationship with Design Traffic

Traffic flow varies from hour to hour, day to day, month to month, and weekday to the weekend (Figures 4.11d1–d3). While AADT provides an annual average daily traffic estimate on a roadway's demand, AADT alone does not give precise enough information for engineers to design a road appropriately.

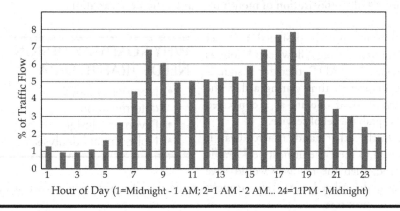

Figure 4.11d1   Illustration of hourly traffic flow variations during an average weekday.

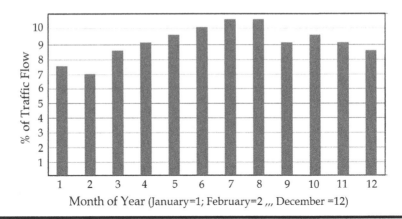

Figure 4.11d2   Illustration of month of year traffic flow pattern.

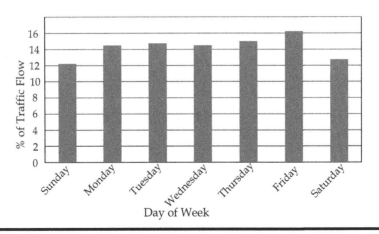

Figure 4.11d3   Illustration of month of year traffic flow variation.

If a roadway is designed to meet the annual average flow condition as expressed in AADT, then for a significant amount of time during a day, the roadway would operate at a congested unacceptable level of service.

On the other hand, if a roadway is designed to meet the utmost peak traffic demand, then the roadway would have many underutilized capacities throughout the day. The situation is then a huge waste of resources.

WHY ROADWAY DESIGN NEEDS TRAFFIC DATA

- Number of lanes determination
- Pavement design
- Noise impact analysis

To balance roadway cost associated with construction, operations, maintenance, and people's desire to travel freely, engineers have adopted the usage of the 30th highest annual hourly traffic volume for a given year as the hourly design volume. Traffic flow data at the 30th highest hourly traffic volume among the 8760 hourly (24 hours/day×365 days/year=8760 hours) observed value in a year is called design hour traffic.

4.4.3.2 K_{30} – Design Hour Factor

Design Hour Factor is the ratio of design hour volume (30th highest hourly volume) to the AADT.

$$K_{30} = \frac{\text{Volume}_{(\text{30th highest hourly volume})}}{\text{AADT}}$$

The 30th highest hourly volume is obtained based on actual continuous traffic monitoring data. For future condition, K_{30} is typically gained by trending historical data.

4.4.3.3 D_{30} – D Factor

D Factor (D_{30}) is the ratio of design hour volume for a given roadway direction to the summation of both directional design hour volumes (DDHVs).

$$D_{30} = \frac{\text{Volume}_{(\text{30th highest hourly volume: one direction})}}{\text{Volume}_{(\text{30th highest hourly volume: both direction})}}$$

D factor shall always be greater than or equal to 50%. If the above computation yields a D factor less than 50%, then the correct D factor is (100% − D Factor).

The 30th highest hourly volume by direction is obtained from historical continuous traffic monitoring data. Future D_{30} is obtained by trending historical data.

4.4.3.3.1 Directional Design Hour Volume

DDHV is the traffic volume engineers use for determining the number of lanes a roadway needs in order to achieve a certain performance standard (e.g., level of service). It can be computed with the equation below.

$$\text{DDHV} = \text{AADT}_{\text{future}} D_{30} K_{30}$$

DDHV: directional design hour volume

4.4.3.4 Number of Lanes Need Estimation

The number of lanes needed for a given roadway is the most critical data element. It can be computed based on the equation below.

$$N_{(\text{lane–needed})} = \frac{\text{DDHV}}{SF_i}$$

where

$N_{(\text{lane–needed})}$: number of lanes needed to achieve the level of service desired

SF_i: maximum practical service flow rate for the level of service desired (can be obtained from standard lookup tables or computed).

4.4.3.4.1 Level of Serves

A road is designed to achieve certain performance expectations. Performance quantification as related to travel reliability and congestion is often characterized by the concept of Level of Service (LOS) prescribed in the Transportation Research Board's Highway Capacity Manual (HCM). The LOS is an indicator of vehicle flow condition, travel reliability, and travel time delay for roads and intersections.

> **SOURCES OF DESIGN TRAFFIC DATA**
>
> - Planning phase travel demand modeling result
> - Refined corridor, subregional, or road specific modeling
> - Field observations
> - Trending analysis

The LOS determination is based on either travel density or travel delay for links and intersections, respectively. The LOS levels include A, B, C, D, E, and F. The LOS A represents free flow with the full ability for free maneuvering. LOS C represents an operating condition at a roadway's capacity. And LOS F characterizes a completely stop-and-go condition where reliability does not exist. For intersections, the LOS is measured with delays as related to travelers' expectations and acceptance.

The analysis of the LOS for freeways and arterial roadways is based on the maximum practical service flow rate (SF_i) (capacity), as outlined below.

$$SF_i = MSF_i \times f_w \times f_{hv} \times f_p$$

where

SF_i: maximum practical service flow rate per lane under a given set of condition (capacity)

MSF_i: maximum service flow rate per lane under ideal condition

f_w: adjustment factor for lane width and lateral clearance

f_{hv}: adjustment factor for heavy vehicle effect

f_p: adjustment factor for senior and student drivers.

The above procedure enables the computation of the actual maximum service flow rate (capacity) per lane under various conditions. The actual maximum flow capacity of a lane should always be less than the maximum flow rate under the ideal state. Adjustment factors reduce a roadway's flowing capacity from the ideal state. For example, if a road's lane is only 9.5 feet wide vs. the standard 12 feet, drivers will be much more constrained and move at a slower speed, reducing the road's vehicle carrying capacity from the ideal situation. These adjustment factors are provided in the HCM manual.

4.4.3.5 Percentages of Medium Truck and Combination Truck

Two more traffic parameters, in addition to the AADT, are needed during the PE/ EIE stage. These two additional traffic data items are medium- and heavy-truck percentages among all vehicles. Truck percentages are required for both pavement design and highway noise impact analysis.

4.4.3.6 Generalized Level of Service Lookup Table

To simplify the LOS computation, generalized analysis can be carried out with default parameters. Results from such generalized analysis are often adequate for planning usage. Table 4.1 illustrates a generalized LOS lookup table produced by the Florida Department of Transportation applicable to Florida Freeways. As shown in the table, for a six-lane freeway located in an urbanized area, if the highway has an AADT greater than 68,100 but less than 93,000, then the level of service is B, which is a very acceptable travel condition in terms of travel maneuverability and freedom.

DESIGN TRAFFIC INCLUDES:

- AADT
- D_{30}
- K_{30}
- Percentages of medium truck and heavy trucks

DESIGN TRAFFIC COVERAGE

- Existing condition (current year)
- Opening year with and without project scenarios
- Design year (20–30 years into the future) with and without project scenarios.

Table 4.1 Generalized Level of Service Data for Florida Highways

Lanes	B	C	D	E
AADT for Core Urbanized Freeway LOS				
4	47,400	64,000	77,900	84,600
6	69,900	95,200	116,600	130,600
8	92,500	126,400	154,300	176,600
10	115,100	159,700	194,500	222,700
12	162,400	216,700	256,600	268,900
AADT for Urbanized Freeway LOS				
4	45,800	61,500	74,400	79,900
6	68,100	93,000	111,800	123,300
8	91,500	123,500	148,700	166,800
10	114,800	156,000	187,100	210,300

Source: Courtesy of the Florida DOT Level of Service Handbook, www.fdot.gov.

4.4.4 Design Speed

Speed is one of the most prominent features of a highway. While posted legal speed limits are for drivers to abide by, nevertheless, a driver operates his or her vehicle at a speed he or she perceives as comfortable and safe. This behavior leads to a wide range of travel speeds on a segment of the same highway at the same time. Field observations show that during free-flow conditions, the 85th percentile of all observed speeds typically matches the posted speed. For this reason, roadways should have adequate geometric features to accommodate safety needs at a minimum of this 85th percentile speed.

The speed selected for designing appropriate geometric features (e.g., horizontal curves, vertical curves, and stopping sight distances) for a roadway is called design speed.

Roadway functional class and area type (urban vs. rural) are typically determined prior to design speed determination. Conditions imposed by roadway functional class and area type limit engineers to a narrow window of final design speed selection.

During the speed determination process, it is important to be realistic on driver expectations for highway speed. For interstate and other freeways, design speed should be as high as feasible.

4.4.5 Access Management

A highway enables movements of both people and goods from one location to another. Its effectiveness depends significantly on how vehicles get on and off it. The control of how vehicles get on and off a highway is called highway access management.

States typically have highway access control statutes for both property owners and state highway agencies to follow. Design engineers must consider all access control issues carefully given they are often controversial.

The FHWA classifies all vehicles into 13 categories. Axle spacing and tire arrangement are the basis for the classification.

4.4.6 Pedestrian and Bicyclist

Walking and bicycling are the most fundamental modes of transportation and are often under-analyzed during highway design. Design engineers need to give proper consideration in determining the need for sidewalks, bicycle lanes, and related safety issues for all modes.

4.5 Preliminary Engineering Analysis and Design

The goal of PE design is to develop an engineering solution that meets the transportation need while avoiding and minimizing environmental impacts.

4.5.1 Horizontal Alignment

A road's physical location on the Earth's surface, as defined by its latitude and longitude data, is called a road's horizontal alignment. A road's horizontal alignment determines where the road is in reference to other landmarks.

A road's horizontal alignment has the most impact and influence on environments and communities.

For a new highway, the horizontal alignment determines where the highway will be, which neighborhood it will go to, and which part of the city it will serve.

For a road-widening project, the horizontal alignment identifies where the additional needed right of way will be acquired. The horizontal alignment totally controls decisions as related to acquiring the needed right of way from left side, right side, or both sides of the existing road.

Figure 4.12a illustrates a 6015 feet segment of the horizontal alignment (from STA 167+95 to STA 1207+80) representing alternatives 8, 14, and 15 for the proposed new Florida Gulf Coast Parkway. The legend panel at the bottom left corner indicates that a white-lines represent property boundaries, dashed grey-lines represent proposed right of way line, and light grey shaded areas are wetland. The alignment shows the bridge has two tangent segments, and these two tangent segments are connected by a simple circular curve starting from STA 211+ 42.08 (PC) to STA 230+62.09 (PT). The curve has a total length of 1920.01 feet.

The proposed highway will need additional right of way from the west side of the existing road as indicated by the dashed grey proposed right of way (ROW) boundary. No other land is needed on the east side of the existing roadway (the existing white property boundary coincides with the proposed grey ROW line).

The alignment also crosses a narrow piece of wetland between STA 220 and STA 224. Wetland impact evaluation needs to be performed.

Horizontal alignment provides boundary information for:

■ Right of way need and relocation assessment
■ Wetland encroachment assessment
■ Wildlife and habitat area encroachment assessment
■ Historical and archaeological site encroachment assessment
■ Floodplain encroachment assessment
■ Highway noise impact assessment
■ Access need assessment
■ Neighborhood cohesion analysis

4.5.2 Vertical Alignment

How a road negotiates hills and valleys along its horizontal alignment is called vertical alignment. Vertical alignment determines how flat or steep a road is and how high or how low a roadway is as compared with its abutting lands.

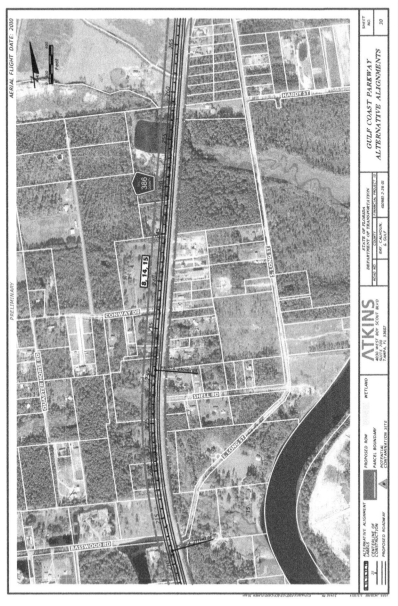

Figure 4.12a Illustration of a horizontal alignment. (Courtesy of Florida Department of Transportation, www.fdot.gov.)

Figure 4.12b shows a vertical alignment for the proposed State Road 64 Intercoastal Waterway bridge. The horizontal grid labeled with the station information characterizes the linear distance. The vertical grid offers elevation information. The scales used are different.

Grey-lines represent the proposed bridge profile. The actual bridge begins at STA 119+22.99 and ends at STA 153+74.64, a distance of 3451.64 feet. The 3451.64 feet bridge has a 784 feet long vertical curve with an up/down 4% grade. While the top grey-line of the bridge profile shows the actual pavement surface elevation, the bottom grey-line indicates the superstructure's bottom elevation. The distance between the bottom and top grey-lines is the superstructure's thickness, which is 8 feet as indicated. Identifying the bottom grey-line is critical as it affects the vertical clearance for water bore traffic (e.g., ships). Here the vertical clearance is 65′ from mean high water elevation per U.S. Coast Guard permitting specification.

At the bridge's left end, an 850 feet vertical curve is used to connect the bridge to the touchdown roadway. On the bridge's right end, a 1,400 feet long vertical curve is used to transition the bridge down to the touchdown road. The vertical alignment extends all the way to other crossing roadways on both ends. This is very critical as impacts from the proposed bridge to other roadways and connecting streets must be fully understood.

The proposed vertical alignment also shows the existing profile (dashed light grey). The existing profile is much lower, with a channel width of 131.21 feet.

Also, water depth information, as depicted by the mean high-water elevation (black line) and the ground line (dashed black line), are also provided.

Vertical alignment offers data and information on:

- Earth fill and cut quantity estimation
- How to tie down a roadway with abutting roadways and driveways
- Guardrail and other roadside feature needs assessment
- Fore-sloping/back-sloping and construction limit determination
- Potential highway noise abatement wall location and height evaluation
- Bridge vs. causeway in dealing with wetland and wildlife crossing
- Underpasses for potential wildlife crossing analysis
- Various vertical clearance needs analysis

4.5.3 Typical Sections

A typical section developed during PE offers not only an engineering solution but also a great public involvement tool due to its ease of understanding. A typical section offers a clear illustration on:

- The right of way (width) needed to accommodate a proposed roadway.
- What a proposed roadway looks like – an arrangement of traveled lanes, bicycle lanes, sidewalks, and other elements.

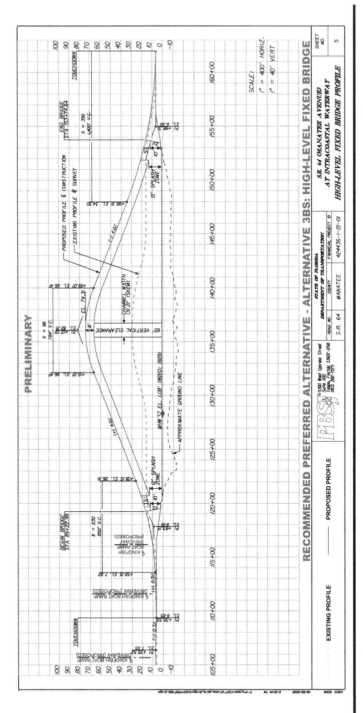

Figure 4.12b **Illustration of a vertical alignment. (Courtesy of Florida Department of Transportation, www.fdot.gov.)**

■ Roadway components to accommodate different constraints (e.g., guardrail with a steep fore slope vs. a wider road with a flatter fore slope).

4.5.4 Design Alternatives

Design alternatives are the derivatives of three key elements – horizontal alignment, vertical alignment, and typical sections. The objective of developing alternatives is to minimize environmental impact and control project costs while fulfilling transportation needs and gaining public support.

4.5.5 Environmental Impact Evaluation

Design engineers begin a design with a wide range of known constraint factors such as (a) 100-year floodplain locations, (b) known wetland areas, (c) known hazardous material locations (e.g., landfill), (d) known wildlife and habitat areas, (e) locations of businesses, (f) residential buildings and homes, (g) known archaeological sites and historic buildings or districts, and (h) both present and future land-use plans.

Once design alternatives are devised, other professionals, such as the right of way agents, relocation specialists, wetland scientists, noise specialists, historians, and archeologists conduct environmental assessments for each alternative. Based on a comprehensive evaluation, design engineers further compare alternatives, modify alternatives, devise new alternatives, and select a final preferred alternative.

4.6 Engineering Product and Deliverable

A PE report is a deliverable documenting all the relevant engineering analysis and engineering decisions made. The topics listed below are typically covered in the PE report.

a. Need for Improvement or Project
  1. Capacity
  2. Safety
  3. Social and economic demand
  4. Consistency with a region's and a state's long-range plan
b. Existing Conditions
  1. Functional classification
  2. Typical section
  3. Pedestrian and bicycle facilities
  4. Drainage
  5. Right of way
  6. Horizontal alignment

7. Vertical alignment
8. Geotechnical data
9. Pavement
10. Intersections
11 Crash data
12. Utilities
13. Structures
14. Railroad facilities
15. Environmental characteristics and land use

c. Design Controls and Specifications
d. Traffic
1. Existing traffic condition
2. Future traffic projections
3. Traffic projection assumption
4. Multimodal consideration

e. Preliminary Design Analysis
1. Design traffic
2. Safety
3. Typical sections
4. Intersection concepts and signal analysis
5. Horizontal and vertical alignments
6. Drainage
7. Right of way
8. Relocation
9. Cost – R/W, relocation, design, construction, recycling of salvageable material
10. Economic and community development
11. User benefits
12. Environmental impacts
13. Utility impacts
14. Traffic control plan during construction
15. Public involvement results
16. Access management
17. Lighting
18. Aesthetics and landscape

f. Alternative Analysis
g. Commitment to the Public and Resource Agencies

Commitments made to the public during public involvement (e.g., to build a noise wall for a given neighborhood) and resource agencies for securing environmental permits (e.g., an underwater construction technique) need to be carried out in final design and construction.

4.7 Next Step

PE develops different engineering solutions iteratively, as environmental impact evaluation reveals new issues. The goal is to avoid and minimize environmental impacts, meet transportation demands, and conform to budget constraints. The PE design report, along with the corresponding environmental impact evaluation report, is presented to the lead federal agency for review and approval. Upon Federal approval, the final design will continue with the PE design.

4.8 Summary

The concepts of the tangent, horizontal curve, and vertical curve provide the foundation for establishing a roadway's horizontal and vertical alignments. Criteria for both horizontal and vertical curve designs are safety-driven. The concept of stopping sight distance serves as one of the critical design control parameters.

The formula:

$$\frac{V^2}{Rg} = \frac{\tan \beta + f}{1 - f \tan \beta}$$

provides the fundamental theoretical linkages among speed, roadway curvature, and superelevation. For practical designs, due to numerous uncertainties associated with some of these parameters, an alternative equation prescribed by AASHTO listed below is used.

$$R_{\min} = \frac{V^2}{15\left(0.01 \, e_{\max} + f_{\max}\right)}$$

Vertical curve design is straightforward with its equal tangent parabolic formula. The determination of the lowest or the highest elevation point is significant because of both drainage (stormwater always runs to the lowest point) and vertical clearance determination needs.

Design traffic data typically covers three different periods. These are current traffic conditions, open-year traffic conditions for both "with" and "without" the proposed project scenarios, and design year for both "with" and "without" the proposed project scenarios. Along with the traffic data, the level of service information, which is a measure of how a road meets the need of drivers, is also provided.

Occasionally, one of the design control's standard specifications may not be able to be met for a variety of reasons. In such cases, a process known as design exception shall be followed. The design exception process allows a systematic analysis of how such an exception may impact other issues and help to generate a comprehensive assessment of its impacts. Only after gaining approval by following the established agency procedure, a design exception can then be implemented during the actual design.

4.9 Discussions

1. Preliminary transportation engineering design offers a footprint for environmental impact evaluation. Environmental impact evaluation provides information to improve and enhance the engineering design. Teamwork is very critical in carrying out a project successfully. Discuss how your style fits in such a teamwork setting.
2. Traffic needs and safety concerns are key drivers for highway projects. Discuss specific issues and factors involved in gathering traffic needs and safety data.

4.10 Exercises

1. Preliminary engineering (PE) provides different horizontal and vertical alignments for a project to support environmental impact evaluation (EIE). The PE and EIE support each other with the goal of avoiding and minimizing environmental impacts. True or False.
2. A roadway plan view is an orthographic projection of a 3-D road to a 2-D horizontal plane. In other words, a plan view is a view from directly above. True or False.
3. A roadway's profile shows how hilly or flat a road is. True or False.
4. A station is a measuring unit used in highway design. One station equals 100 feet. True or False.
5. Guardrails used for highway safety purpose have two parts: the rail body and rail end. True or False.
6. Overhead variable message boards installed on interstate highways are typically controlled remotely by a Traffic Management Center. True or False.
7. Driveway connections are affected by a roadways' vertical alignment (profile). True or False.
8. The distance from Station 80+12 to 89+25 is 913 feet. True or False.
9. The tangent segment of a road is the section that is not part of a horizontal curve. True or False.
10. A circular curve can be used to link two tangent roadways segments while providing a smooth turning path for vehicles. True or False.
11. The severity of a horizontal curve, commonly known as the sharpness of the curve, can be characterized by either its radius or its curvature. True or False.
12. A 3° curve has a much sharper turn (smaller turning radius) than a 2° curve. True or False.

13. An east-west roadway turns 45° to the southeast direction, as illustrated below. The roadway's curvature is 45°. True or False.

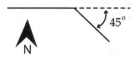

14. What is the curvature in degrees for a 2680-feet radius horizontal circular curve?

15. A broken back circular curve layout can be used in the high-speed design. True or False.

16. A reverse circular curve layout can be used with local low-speed roads. True or False.

17. In the field, a person can easily use a flexible steel measuring tape to measure the length of a curve with good accuracy. True or False.

18. The intersection of two tangent roads in a horizontal plane is called the point of intersection (*PI*). It is possible that a *PI* resides in a lake or a body of water. True or False.

19. A parabolic vertical curve connects two different grade roadway segments providing a constant turning rate that meets drivers' natural body intuition. True or False.

20. A roadway vertical curve always takes the form, which can be characterized by a circular curve formula. True or False.

21. A vertical curve takes shape characterized by a parabolic formula. True or False.

22. A vertical curve always has a maximum or minimum elevation point along the curve. True or False.

23. There are two types of basic vertical curve layouts: sag and crest. True or False.

24. Sight distance is the distance ahead of a driver where the driver can see. True or False.

25. Sight distance is affected by a lot of factors. Identify all applicable ones listed below.
    a. Fog
    b. Height of the driver's eye
    c. Height of an object ahead of the driver
    d. Vehicles in front of the driver's vehicle
    e. Dirty on a windshield
    f. Music played in driver's vehicle
    g. Passengers in a car
    h. The horizontal curve of a road
    i. Vertical curve of a road

26. Stopping sight distance is the distance it takes to completely stop a vehicle after a driver senses the need to stop. True or False.

27. Superelevation is used to facilitate the easiness of vehicle turning associated with horizontal curves. Identify four parameters design engineers need to pay attention to from the list below.
    a. Number of vehicle axles
    b. Gross vehicle weight
    c. Horizontal curve radius
    d. Transversal tire and pavement friction factor
    e. Superelevation
    f. Design speed
    g. Animal crossing
28. The superelevation typically does not get higher than 12%. What's the concern for a too steep superelevation embankment?
29. Why does a roadway need a minimum grade vs. a completely flat vertical alignment?
30. For practical design purposes, the height of the eyesight is set at 3.5 feet, and the height of an object on the road is set to be 2.0 feet. Is the above statement true?
31. Truck drivers' eyesight is much higher than 3.5 feet. This extra elevation provides truck drivers' ability to see further ahead. This extra sight distance compensates the extra stopping distance trucks may need. Is the above statement correct?
32. Label the cross section illustrated below for all missing elements.

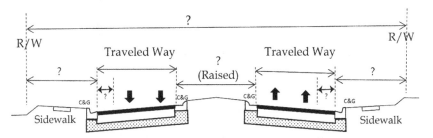

33. Label the typical section illustrated below for all missing elements.

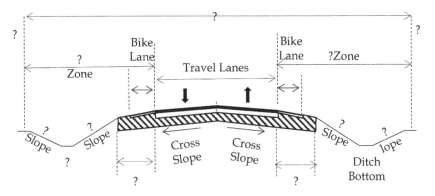

34. The Federal Highway Administration classifies all highway vehicles into 13 vehicle classes. True or False.
35. Traffic data used for highway design includes data covering the current year, design year with the project, and design year without the project. True or False.
36. Future design traffic data for a preliminary engineering and NEPA project can be obtained from many different methods. These methods may range from simple linear extrapolation of historical data to sophisticated travel demand modeling. True or False.
37. D factor in traffic engineering is known as a directional factor during peak hour travel. For urban area radial roadways, reported D factors are 50% at peak hour. The data reviewer concludes that the 50% D information reported is most likely wrong. Is the reviewer correct?
38. Design Hour Factor is also known as K_{30} factor. It is the percentage of the 30th highest hourly flow volume over the AADT. The use of K_{30} vs. K_1 or K_{10} is a balancing act between roadway construction cost and the flow of traffic need. Is the above statement correct?
39. Pedestrian and bicycle travel are valid modes for transportation planning purposes. True or False.
40. A roadway's horizontal alignment provides information and basis for the following subject areas. Select all applicable ones.
    a. Location concerning where the roadway will go through
    b. Right of way assessment
    c. Highway noise impact analysis
    d. Bridge vertical clearance
    e. The footprint for wetland analysis
    f. The footprint for wildlife and habitat analysis
    g. Inform property owners whether the roadway will impact them or not from a right of way standpoint
41. A roadway's vertical alignment provides information and basis for the following subject areas. Select all applicable ones.
    a. Construction limit determination
    b. How to tie the edges of roadways to existing grounds at the right of way line
    c. Bridge vertical clearance
    d. Underpass for wildlife crossings and clearance need
    e. Earth cut and fill needs
    f. The footprint for wildlife and habitat analysis
    g. Guardrail need assessment
42. A driver was traveling on a level country road with a speed of 68 mph. All a sudden, the driver noticed that there was a large piece of rock on the road ahead. He slammed on the brake. The vehicle stopped just 2 feet away from crashing onto the rock. How far did the car may have traveled after the driver saw the rock?

43. The spiral curve layout offers a very even and smooth transition from a tangent to a curve. And the spiral curve layout is used extensively for railway design. Why is it not popular with highways?

44. Residents at the Green Valley community have complained to the City transportation department that motorists are speeding through their neighborhood. The road involved has a posted speed of 35 mph. The City's traffic operation technician visited the site and collected over 300 vehicle speeds in 2 hours. The speed data shows that the 85th percentile speed is 31 mph. What does this 85th percentile speed mean?

45. For a planned highway, the estimated AADT for the design year is 246,600. The D-factor is 65%, the estimated K_{30} is 11%. The service flow rate of 1,810 vehicles per hour per lane for a LOS C is desired. How many lanes are needed if LOS C is the goal?

46. Preliminary engineering (PE) report covers both current facility conditions and proposed concepts and designs. Make a list of key elements contained in a typical PE report.

Bibliography

Highway Capacity Manual, Edition 6: A Guide for Multimodal Mobility Analysis, Transportation Research Board (TRB).

A Policy on Geometric Design for Highways and Streets, 2018, American Association of State Highway and Transportation Officials (AASHTO Greenbook).

The Project Development Process Manual, Ohio Department of Transportation, http://www.dot.state.oh.us/Divisions/Planning/Environment/manuals_guidance/Documents.

Chapter 5

Environmental Analysis

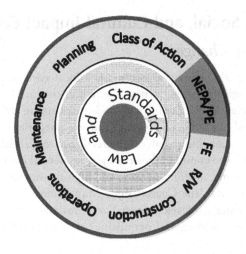

While the National Environmental Policy Act (NEPA) declares a national policy to "...encourage productive and enjoyable harmony between man and his environment; to promote efforts which will prevent or eliminate damage to the environment and biosphere and stimulate the health and welfare of man; to enrich the understanding of the ecological systems and natural resources...," NEPA lacks specificities on how this law could be implemented in practice. To remediate this, Congress passed a host of procedural, process, and standard focused legislation such as the Clean Water Act (CWA), the Clean Air Act, the Endangered Species Act, the Noise Control Act, the Resources Conservation and Recovery Act, the Farmland Protection Policy Act, etc. Additionally, the Executive branch promulgated a wide range of detailed Federal regulations to carry out the intent of NEPA. The aforementioned legislation and regulations have provided detailed procedures and criteria in carrying out the spirit of NEPA.

Environmental impact evaluation (EIE) is typically grouped into the following three subject areas:

1. human, social, and cultural EIE,
2. natural EIE, and
3. physical EIE.

For each engineering design alternative, design engineers, working with other subject matter experts, perform analysis for the aforementioned subject areas with the goal of avoiding environmental impact and minimizing any impact in cases where avoidance is not feasible. Finally, when impacts are unavoidable, design engineers must work to mitigate these as commitments.

5.1 Human, Social, and Cultural Impact Evaluation

5.1.1 Land-Use Changes

Land-use planning, also known as zoning, is a local government effort to guide development through permitting types of activities that are legally allowed to be carried out in different areas. Zoning and land-use planning are often presented on a map, as illustrated in Figure 2.7. Typical land-use categories include agricultural, natural and conservation, industrial, commercial, and residential. Further divisions of each category are also common. For example, residential land use can be divided into single-family home, medium-density, and high-density residential areas.

On the Federal front, the Coastal Zone Management Act encourages coastal and Great Lakes states to develop coastal zone management programs ensuring that Federal actions (funding and permitting) are consistent with state management plans.

For highway projects, the focus is to ensure that federally funded highway projects are compatible and supportive of local government land-use plans.

Specific consistency and compatibility assessment questions include:

1. Alignment – does a road's alignment lead to a new development that is not compatible with the locally approved land-use plan?
2. Access – does a proposed roadway provide needed access points to an area or region?
3. New development – does a proposed roadway open new development, which is not desired per the locally adopted land-use plan?
4. Engineering compatibility – is the engineering design compatible with both current and future land-use plans? Specific design issues such as curb and gutter vs. open ditch drainage, decorative highway lighting vs. high mask lighting, bike lane and sidewalk, right of way reservation, and others should be considered.

5.1.2 Community Cohesion

While roads and highways provide connections and accesses to people, neighborhoods, cities, and regions, roadways may also separate people, split a neighborhood, or divide a town or city.

Design engineers should pay close attention to how people cross an existing roadway if the proposed action is to widen that roadway. Crossing volume, crossing locations, community demographics (e.g., senior citizens, children), and frequency of crossings should be observed and understood.

Engineers should also realize that crossing a two-lane road with relatively low traffic volume is a lot easier than crossing a four-lane high-volume road. Providing adequate at-grade crossing points and considering overpass or underpass crossings are essential to minimize interruptions to people's lives. When considering an overpass or an underpass, it is important to address accessibility for people with disabilities (e.g., needs of ramps for wheelchairs), vertical clearance needs for overpasses, and potential flooding issues for underpasses.

For any new highway, design engineers should devote attention to preserving neighborhoods. Effort should be given to preserving existing neighborhoods and not splitting them (Figure 5.1).

Figure 5.1   Illustrations of various pedestrian bridges.

If a new highway is a limited-access highway, ensuring that people on either side of the proposed highway can easily cross the highway through an overpass and/or an underpass without major hindrance and inconvenience is a priority. Underpasses and overpasses should accommodate not only pedestrians and passenger vehicles but also farming equipment and agricultural animals. Underpasses or overpasses should be located conveniently for residents.

The bottom line is that a roadway should not be a barrier to people living on either side of it and should not cause major inconvenience to people who live and work in the area.

5.1.3 Aesthetic

Aesthetic refers to the outward appearance or beauty of a physical object. Roads, highways, and bridges are objects of great physical scale. Their outward appearances are often further augmented through their style, material, color, and architecture. Design engineers should take appropriate actions to ensure that the aesthetic of a roadway or bridge is considered.

The U.S. Department of Transportation (DOT) Order 5610.1C requires all Federal transportation projects to consider aesthetic quality. The Federal Highway Administration (FHWA) has issued specific guidelines on how to carry out this order.

Design engineers and all professionals involved should consider:

- Location of a roadway project
- Signs
- Color and patterns of structures and retaining walls
- Trees/lawns/wildflowers and other greeneries
- Lighting, lighting pole, and lamp style

5.1.4 Relocation

Relocation, in the context of any highway project, is related to the acquisition of real property from private citizens or businesses. According to the U.S. Constitution, the property owner must be paid just compensation. The Uniform Relocation Assistance and Real Property Acquisition Policies Act of 1970, commonly referred to as "The Uniform Act," prescribes exactly what Federal agencies or their enabling state or local agencies, such as State transportation departments, must do when acquiring right of way for a Federal or Federal-aid transportation project. Design engineers should be conscious of the following:

1. Avoid and minimize relocation – not only relocation is expensive to the project owner but also causes other issues such as emotional anxiety and financial hardship for relocatees.
2. Avoid isolating any remaining homes or businesses.

3. Avoid splitting a neighborhood if the additional right of way needed cannot be avoided.

4. Ensure that potential relocatees are fully aware of their rights as defined under the Uniform Act and the timeline for the project.

5. Assess potential controversies. Some homeowners and businesses may be eager to sell their properties,

OVERPASS DESIGN

Vertical clearance needs must be fully considered.

UNDERPASS DESIGN

Preventing flooding and having lighting must be fully considered.

and others may be very reluctant. The key is to understand people's positions.

6. Work with the right of way and relocation professionals to estimate both the right of way and the relocation costs for various design alternatives.

State and local governments may have their own unique right of way acquisition laws and procedures. However, for any project where Federal fund is used, or a Federal permit is sought, the Federal Uniform Act must be followed.

5.1.5 Community Services

Community services refer to both physical community service facilities and how people travel to and from these facilities. Community service facilities include public libraries, schools, community centers, hospitals, and fire and rescue stations. When designing a highway, access changes to these facilities must be considered. Specifically, the following issues should be examined and considered:

■ Impacts on bicycling and walking routes to and from these facilities. This includes bicycle lanes, sidewalks, roadway crossings, and other safety and distance issues.

■ Access points for fire and rescue vehicles and potential route changes.

The Federal legislation contained in 23 U.S.C. Section 128 Public Hearings states that all federally funded or permitted projects shall offer public hearings or hearing opportunities. During these hearings, effects and outcomes from different design alternatives are presented by the responsible transportation agency, an opportunity for public commenting is offered, and suggestions from the public are heard, recorded, and considered.

5.1.6 Historic and Archaeological Resources

Section 106 of the National Historic Preservation Act requires that all federally funded, licensed, permitted, or approved projects must be evaluated for their impact and potential impact on historic and archaeological resources.

Figure 5.2   Historic sites or districts.

Section 106 requires the lead Federal agency to consult with the State Historic Preservation Officer (SHPO) on potential adverse effects from a proposed project (Figure 5.2).

One of the most critical steps in the cultural and historic impact assessment is the identification of historic resources. Obviously, structures, districts, landmarks, and others already listed on the National Register of Historic Places are known historic resources. The key is to identify unlisted but potentially eligible properties. Cultural resource specialists (trained professionals in historic and archaeological evaluation) analyze potential archaeological sites, buildings, objects, etc., of 50 years of age or older and determine their potential historic significance under each design alternative.

Impacts on historic and archaeological resources could range from a direct take to limiting and altering access points to such resources.

Strategies for dealing with cultural resource impact range from avoidance, minimization, mitigation, to recordation. Recordation refers to efforts to record specific physical sites, time or historic periods, and communities for education, display, and preservation.

5.1.7 Parks and Recreation Areas

Parks and Recreation Areas are parklands, recreational areas, waterfowl and wildlife refuges, and significant historic sites that are owned by public agencies and open to the public.

Section 4(f) of the U.S. DOT Act requires that any significant use of such an area or a portion of such an area for the purpose of transportation can only proceed when there is no prudent and feasible alternative and that the design of a roadway project should include all possible alternatives to minimize harm to the site.

5.2 Natural Environmental Impact Evaluation

5.2.1 Wetland

Wetlands are areas where water either covers the soil or is near the soil surface all year long or for a significantly long period. Soil, water, plants, and other living organisms acting together create a unique wetland ecosystem.

The unique wetland ecosystem plays a critical role in human society. It acts as a filter for water. When water from high ground flows and drains to a lower elevation basin, it brings sediment, nutrients, and other natural or human-made substances such as fertilizers and pesticides with it. This low-lying basin is the wetland ecosystem. Here microbes absorb nitrogen, sulfur, and other nutrients, break down other compounds, and form the base of the food chain. Further along the food chain, organisms such as fish, amphibians, shellfish, and insects grow and consume these nutrients. Many larger organisms, such as birds, turtles, and alligators, also rely on wetlands for food, water, and shelter.

Section 10 of the Rivers and Harbors Act, Section 404 of the CWA, Emergency Wetlands Resource Acts, and Executive Order 11990 Protection of Wetland, all have statutory requirements to protect wetlands. Unlike many other environmental protection regulations that are only applicable to Federal actions, wetland impact analysis and mitigation are required for all projects, including private projects.

Wetland impact evaluation starts by identifying the type, quantity, quality, and function of the wetlands involved. This preliminary identification often begins with the digital National Wetland Inventory imagery map. By overlaying the wetland inventory map over an engineering design alternative, the quantity (acres) of wetland impacted and types of wetland (e.g., riverine, slope depressional, flat, and fringe) impacted can be analyzed. Additionally, the primary function of the wetland involved and the relative importance of the impacted wetland can be compared with other wetlands in the area (same watershed).

During the alternative development process, coordination with the U.S. Army Corps of Engineer (COE), local district, and a State's environmental resource agency should occur as early as possible. An early buy-in on a highway agency's design concept from the COE and the state's environmental resource agency is critical to ensuring a viable and acceptable alternative for the project (Figure 5.3).

One of the last critical steps in wetland impact analysis is the official wetland boundary determination. The legal boundary defining where a wetland begins and ends is called a jurisdictional boundary. The jurisdictional boundary is determined from consensus work between state highway agency's wetland experts and the U.S.

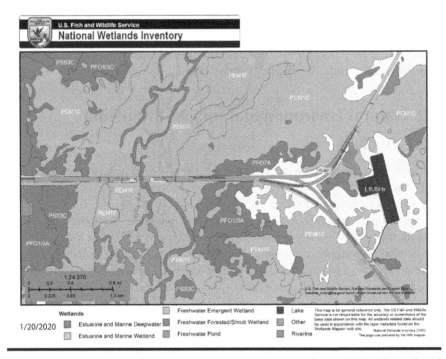

Figure 5.3 Sample wetland inventory map. (Drawn from USFWLS National Wetland Inventory GIS data at https://www.fws.gov/wetlands/Data/Mapper. html.)

Army COE wetland experts in the field through both soil profiles and plant species analysis. Once a boundary is established, the exact amount (acreage) of impact can be computed.

The COE offers voluntary COE wetland delineation training and certification. Those who pass the COE examination are credentialed as the U.S. Army COE-certified wetland delineators.

While design engineers should avoid and minimize wetland impacts, there are situations where wetland impact is unavoidable. In these situations, minimizing impact is practiced, and mitigating the unavoidable impact is carried out.

There are many ways of mitigating unavoidable wetland impact. Creating a new piece of wetland next to the impact one, creating a single large piece of wetland to mitigate all impacted wetland from all projects in the same watershed area, and creating a large contiguous wetland to mitigate all impacts from all projects occurred in a region are all potential mitigation options.

Historically, a single piece of wetland was created next to the impact site (on-site wetland) for mitigation purposes. However, it was discovered that this newly created isolated wetland does not function well pertaining to the wetland functions. The current approach is to create large contiguous wetlands in a region. These large wetlands created for mitigation purposes are called wetland mitigation banks. States or private businesses can voluntarily create (typically restore) large acreage of functional wetlands. These newly created wetlands act like bank deposits (credit). As a deposit (credit), it can be drawn to meet mitigation needs. Also, these deposits can be sold on the marketplace to those who need to mitigate impacts. A purchaser can acquire such wetland instead of creating his or her own wetland.

When a piece of wetland is impacted (filled), it is unlikely that the replacement wetland (mitigation) will be the same size as the original (1:1 ratio). In most cases, for every 1 acre of wetland impacted, more than 1 acre of replacement is needed. This ratio depends on the quality of the impacted wetland. The higher the quality of the impacted wetland, the higher the mitigation ratio is. This ratio is a negotiated result between a State highway agency's wetland permitting professionals and an environmental resource agency's personnel.

On the Federal level, the U.S. Army COE regulates wetland.

For example, a state highway agency's project impacts a total of 2.1 acres of wetland. Through negotiation with environmental resource agencies, it is determined that the mitigation ratio is 1:1.8. The state will need to create $2.1 \times 1.8 = 3.78$ acres of new wetland. The state can purchase this 3.78 acres of wetland from a wetland mitigation bank to satisfy its mitigation obligation.

During the preliminary engineering and EIE stage, wetland impact and mitigation analyses are not the final conclusions given that the design plan is still preliminary. In the Preliminary Engineering Report and the EIE document, a state transportation department commits to mitigating wetland impact as the project is moved forward.

Before construction can begin, a state highway agency must have the wetland impact permit (also known as Section 404 permit) in hand.

5.2.2 Wildlife and Habitat

The Endangered Species Act applies to both public and private actions that will likely jeopardize the continued existence of endangered or threatened biological species or destroy critical habitat for such endangered or threatened species. The U.S. Fish and Wildlife Service and the National Oceanic and Atmospheric Administration (fisheries) share responsibilities in administering this legislation.

Several key definitions used in the analysis of wildlife and habitat, per various Sections of Title 50 Code of Federal Regulation, are listed below.

5.2.2.1 Action Area

Action areas refer to "all areas to be affected directly or indirectly by an action and not merely the immediate area involved in the action." This is very critical as an action area could be outside of a project (right of way) limit.

5.2.2.2 Incidental Take

An Incidental Take is the "take of listed fish or wildlife species that results from, but is not the purpose of, carrying out an otherwise lawful activity conducted by a federal agency or applicant, or contractors working on behalf of the applicant."

During the impact analysis process, early consultation with the U.S. Fish and Wildlife Service ensures an alternative's viability. Additionally, subject matter experts at the Federal and state environmental resource agencies can help to discover issues early on and offer options for highway engineers to consider in order to avoid and minimize impacts.

5.2.3 Floodplains, Stormwater Runoff, and Water Quality

Executive Order 11988, Floodplain Management requires that the construction of all federal and federal-aid facilities, including buildings, roads, and physical objects encroaching upon base floodplain (100-year floodplain), needs to consider alternatives to avoid direct and indirect take of floodplain areas.

U.S. DOT Order 5650.2 Floodplain Management and Protection and Federal-Aid Policy Guide 23 CFR 650A outlines specific criteria and approaches in dealing with floodplain issues.

Per 23 CFR 650A, "It is the policy of the FHWA:

a. To encourage a broad and unified effort to prevent uneconomic, hazardous or incompatible use and development of the nation's floodplains.
b. To avoid longitudinal encroachment, where practicable.
c. To avoid significant encroachments where practicable.
d. To minimize impacts from highway agency actions that adversely affect base floodplains.
e. To restore and preserve the natural and beneficial floodplain values that are adversely impacted by highway agency actions.
f. To avoid the support of incompatible floodplain development.
g. To be consistent with the intent of the Standards and Criteria of the National Flood Insurance Program, where appropriate.
h. To incorporate 'A Unified National Program for Floodplain Management' of the Water Resources Council into the FHWA procedures."

The intent of these laws and regulations is to avoid and minimize encroachment from highways and highway-supported land development that reduces stormwater storage capacity and increases water surface elevations within 100-year floodplains.

5.2.3.1 Stormwater Runoff Quantity (Flooding) Control

Pavement surfaces are impervious. Rain or other precipitation falling on a road will not be able to permeate to the underlying soil. Instead, the water runs off the

Figure 5.4a   Stormwater detention pond illustration.

pavement surface in greater quantity (due to a lack of vegetation interception and soil adsorption) with faster speed (roadway cross scope to keep the traveled way dry) than it would under the natural condition. This phenomenon may cause flooding of adjacent properties (Figure 5.4a).

To maintain the natural stormwater drainage pattern and prevent flooding, runoff is channeled into stormwater retention or detention ponds first. From there, discharging the runoff is controlled at a rate not to exceed its "natural state" rate. The "natural state rate," also called "pre" rate refers to the runoff rate in the absence of any roadway facility. The guiding principle is that the post-discharge rate (with the presence of a roadway system) shall not exceed the natural state pre-discharge rate.

Stormwater pond sizing, channeling, and discharge facilities are designed by drainage engineers (hydrology).

5.2.3.2 Stormwater Water Quality Control

5.2.3.2.1 Construction Activity Related

The CWA prohibits discharging pollutants through a point source into the "water of the United States" unless the discharge is permitted under a National Pollutant Discharge Elimination System (NPDES) permit.

The NPDES program is managed by the U.S. EPA and is often delegated to State environmental resource regulatory agencies for permit administration. The Stormwater NPDES program applies to all outdoor construction activities. This includes land clearing, soil grading, and soil excavating in areas of soil stockpiling, soil borrowing, material storage, and equipment storage. Highway construction falls within these specified activities.

A construction activity owner (an activity owner – e.g., a contractor), not the project owner (e.g., a state DOT), is generally responsible for obtaining the NPDES permit from a state's environmental regulatory agency. A permit represents written permission from a regulatory agency allowing a construction company to conduct fieldwork under the condition outlined in the permit. Construction companies comply with all permitting conditions by developing and implementing stormwater pollution prevention plans as outlined below.

1. Identifying potential sources of stormwater pollutants including sediment, oil, and grease from storage and spill, trash, sanitary and septic, and others.
2. Identifying and implementing good practices, including isolation and filtration, and good housekeeping conservation techniques to prevent pollution.
3. Installing erosion control material, monitoring compliance, reporting and remediating violations.

5.2.3.2.2 Operations Related

Once a road opens to traffic, a wide range of contaminants ranging from leaked engine oil, vehicle and cargo spills, dust and particles from tire wear and tear, dust from brake wear, to debris from roadway surface deterioration accumulate on the roadway surface. When precipitation falls on a road, these accumulated materials are washed away, potentially polluting receiving water bodies (Figure 5.4b).

To protect receiving water quality, stormwater runoff from a road is required to be retained for a certain amount of time before it can be discharged into lakes, rivers, or any other aquifer. This retention enables the settlement of various suspended

Figure 5.4b   Illustration of leaked oil on roadway surface.

particulate materials. Stormwater retention ponds are designed to meet both water quantity (flood) and water quality needs.

5.2.4 Farmland

The Farmland Protection Policy Act (FPPA) is created to minimize, through projects and programs permitted or

HIGHWAY STORMWATER MANAGEMENT IS TO PREVENT:

- potential flooding of adjacent properties and
- polluting receiving waterbodies.

funded by the Federal government, the conversion of farmland to a nonagricultural use. The FPPA protection applies to farmlands even if such lands are not in active use as cropland. FPPA lands can be forest land, pastureland, cropland, or any other lands, as long as they are not water or urban built-up land.

The Natural Resources Conservation Service (NRCS) of the U.S. Department of Agriculture is responsible for ensuring that FPPA is implemented.

5.3 Physical Environmental Impact Evaluation

5.3.1 Highway Noise

While vehicles provide the means for people to move from one place to another, vehicles also generate noise. When a home is close enough to a highway, the highway-related noise level at home may become unacceptable and unhealthy for its residents. To protect residents from harmful noise, Congress passed the Noise Control Act, which directed the FHWA to establish standards for roadway noise control.

The FHWA developed and codified the "Procedures for Abatement of Highway Traffic Noise and Construction Noise" in 23 CFR Part 772. Table 5.1 lists highway noise abatement criteria established in the aforementioned Federal regulation.

Noise abatement criteria for different locations vary. The type of activity performed (Category A, B, C) at a site and whether the activity is exterior or interior orientated determine the applicable noise abatement criterion.

23 CFR Part 772 is only applicable to Federal-funded new alignment highway projects or capacity improvement projects (roadway widening projects) where additional through lanes are added.

State governments establish state-specific regulations in dealing with other scenarios (e.g., existing highway noise impact) related to highway noise abatement.

As highways are widened, roads become closer to homes. Therefore, highway noise impact analysis and abatement have become increasingly challenging for highway agencies. State DOTs have developed and adopted a wide range of noise abatement measures, including the highly visible noise abatement walls, as illustrated in Figure 5.5a. Highway noise analysts must understand not only the science

Table 5.1 Noise Abatement Criteria Per 23CFR772 (Hourly A-Weighted Sound Level[a] decibels)

Activity Category	Activity Criteria[b]		Evaluation Location	Activity Description
	Leq(h)	L10(h)		
A	57	60	Exterior	Lands on which serenity and quiet are of extraordinary significance and serve an important public need and where the preservation of those qualities is essential if the area is to continue to serve its intended purpose.
B[c]	67	70	Exterior	Residential
C[c]	67	70	Exterior	Active sport areas, amphitheaters, auditoriums, campgrounds, cemeteries, day care centers, hospitals, libraries, medical facilities, parks, picnic areas, places of worship, playgrounds, public meeting rooms, public or nonprofit institutional structures, radio studios, recording studios, recreation areas, Section 4(f) sites, schools, television studios, trails, and trail crossings
D	52	55	Interior	Auditoriums, day care centers, hospitals, libraries, medical facilities, places of worship, public meeting rooms, public or nonprofit institutional structures, radio studios, recording studios, schools, and television studios
E[c]	72	75	Exterior	Hotels, motels, offices, restaurants/bars, and other developed lands, properties or activities not included in A–D or F.

(Continued)

Table 5.1 (*Continued*)   Noise Abatement Criteria Per 23CFR772 (Hourly A-Weighted Sound Level[a] decibels)

Activity Category	Activity Criteria[b]		Evaluation Location	Activity Description
	Leq(h)	*L10(h)*		
F	–	–	–	Agriculture, airports, bus yards, emergency services, industrial, logging, maintenance facilities, manufacturing, mining, rail yards, retail facilities, shipyards, utilities (water resources, water treatment, electrical), and warehousing
G	–	–	–	Undeveloped lands that are not permitted

Sources: 23 CFR Part 772, https://www.gpo.gov/fdsys/.

[a] Either Leq(h) or L10(h) (but not both) may be used on a project.
[b] The Leq(h) and L10(h) Activity Criteria values are for impact determination only, and are not design standards for noise abatement measures.
[c] Includes undeveloped lands permitted for this activity category.

• Using concrete walls for highway noise abatement is one of the most widely adopted approaches.

• These pictures illustrate a) wall styles, b) surface textures, c) colors, and d) vegetation coverages to meet the desires of affected residents.

Figure 5.5a   Illustrations of noise abatement walls.

of noise modeling and measurement, but also relevant policy, regulations, and laws. Additionally, analysts must be able to communicate effectively with the public both orally and in writing.

5.3.1.1 Basic Noise Science

Vibrations cause changes in air pressure. These pressure changes create a series of pressure waves. When a pressure wave travels through air and reaches people's eardrums, people sense the pressure change and describe it as sound. A sound is a composite of different pressure waves. Undesired and annoyance sound is termed noise.

5.3.1.1.1 Sound Frequency (f)

Frequency (f) is a measure of the number of vibrations per second and has a unit of hertz (Hz). One Hz is equal to one event cycle per second.

Human hearing perceives sound frequency as pitch (e.g., a high-pitch violin or a low-pitch trombone). Humans can decipher frequencies between approximately 20 Hz and 20,000 Hz. As people age, the ability to hear high-frequency sound tends to decline.

5.3.1.1.2 Sound Speed (c)

The speed at which sound travels (c) is primarily affected by the density and the compressibility of its traveled medium. For highway-related noise, the travel medium is air. The sound speed in the air is approximately 1,130 feet per second.

5.3.1.1.3 Sound Wavelength (λ)

Wavelength (λ) is the distance traveled by a sound wave during one cycle. Sound waves audible to humans range from approximately 1 inch to 55 feet in length.

5.3.1.1.4 Relationship

Frequency, wavelength, and speed of sound waves are linked by the equation $c = f \times \lambda$. Sound speed is the mathematical product of its frequency and wavelength.

5.3.1.1.5 Octave Bands

An octave band divides the entire frequency range into smaller segments called octaves. The method for this division is that the higher-frequency octave segment is twice that of the lowest frequency. It quantifies effective frequencies without looking at each frequency one at a time. One of the most common octave frequency bands is 31 Hz, 63 Hz, 125 Hz, 250 Hz, 500 Hz, 1 kHz, 2 kHz, 4 kHz, 8 kHz, and 16 kHz.

5.3.1.1.6 Sound Energy and Pressure

Sound power is the amount of energy per unit of time that is emitted from a source in the form of sound waves. Linking sound power and sound pressure is critical in building modeling tools to predict noise levels under various conditions.

While it is not possible to measure acoustic energy directly, pressure changes derived from vibrations can be detected easily and precisely. Acoustic energy is proportional to sound pressure squared. The sound pressure threshold for hearing is 20 micro-pascals (1,000 Hz).

5.3.1.1.7 Sound Measuring Unit – Decibels

Decibels (dB) is the unit used to describe the intensity of sound. The mathematical equation is:

$$dB = 20 \log \frac{P_2}{P_0}$$

P_2: pressure in pascals (Pa)
P_0: reference pressure 0.00002 Pa (20 micro-pascals)

Decibel is on a logarithmic scale. A small change in the dB level means a large change of energy. When P_2 is at its lowest human decipherable pressure of 0.00002 Pa, the noise level is equal to zero.

$$dB = 20 \log \frac{0.000002}{0.000002} = 0$$

5.3.1.1.8 Loudness of Sound

Loudness is how human ears sense sound. Sound loudness depends on both the acoustic pressure and frequency. Three internationally standardized sound frequency weighting methods known as A, C, and Z frequency weightings are used to characterize sound loudness.

A-weighting (dBA) deemphasizes low-frequency waves. Human ears are less sensitive to both low and very high frequencies. A-weighting is commonly used for environmental noise measurement.

> Lowest human decipherable sound pressure is approximately 20 micro-pascals.

C-weighting (dBC) simulates human ears' frequency sensitivity at the very high noise level. The C-weighting scale is very flat as compared with A-weighting because it includes a greater range of low-frequency sounds than the A scale.

Both A and C weightings are attempts to simulate how human ears react to sound.

Z-weighting using the same weight for all frequencies, and it is also referred to as no frequency weighting.

Highway-related noise criteria shown in Table 5.1 are based on based on A-weighting.

Noise measuring equipment for highway field noise monitoring should be equipped with the A-scale weighting function.

WHY SOUND MEASUREMENT INSTRUMENTS HAVE WEIGHTING DEPLOYED

Human ears react differently to different sound frequencies. Weighting is an attempt to simulate human ears' reaction to a composite sound wave.

5.3.1.1.9 Leq and L10 Sound Levels

Leq is an equivalent sound level for a specific period. A 1-hour Leq is an equivalent (average) constant sound level for a whole hour.

L10 sound level is the sound level exceeding 10% of the time during a specific period. And 1-hour L10 is the sounding level exceeding 10% of the time during a whole hour.

5.3.1.1.10 Insertion Loss

Insertion Loss associated with highway noise analysis refers to noise level reduction resulting from a noise wall, a berm, or other installed barriers or devices. For example, with a project, it is determined that a 12.6 feet high concrete noise abatement wall can reduce 8.2 dBA of noise. The 8.2 dBA reduction is called "Insertion Loss."

5.3.1.1.11 Sound Transmission in the Field

Sound or noise impact analysis is the assessment of sound propagation from a source to a receiver. A receiver is a location representing an outdoor area where there are human activities. The most commonly identified outdoor activity area in a highway setting is a home's backyard illustrated in Figure 5.5b. The centroid (x, y, z) of the backyard is typically used to represent the impact location. The sound level at the centroid site is compared to the noise impact criteria listed in Table 5.1.

Figure 5.5b illustrates the geospatial layout among sound source (road), terrain, sound wall, and a receiver.

As vehicle noise propagates toward a house, the sound is reduced from (a) geometric attenuation, (b) atmospheric absorption, (c) terrain ground surface bouncing, and (d) barrier blockage if one is in existence.

Sound geometric attenuation refers to the reduction in sound pressure caused by the spread of the radiated sound energy over a sphere of increasing area $(4\pi r^2)$. The area of spherical wave front increases in proportion to the square of the distance r (radius of the sphere) from the source. The resulting sound pressure level decreases

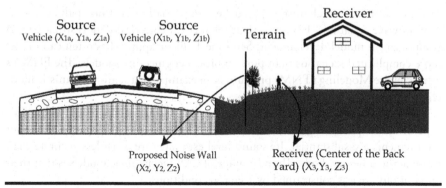

Figure 5.5b   Noise transmission from source to receiver illustration.

at a rate inversely proportional to r^2. This inversed squared relationship is why distance itself is a very effective measure to mitigate noise impact.

Atmospheric absorption attenuation is the conversion of molecular oscillation energy to heat due to the viscous frictional movement of air molecules.

Terrain ground surface bouncing covers both ground absorption and reflection of sound waves. This ground refection leads to sound wave direction changes and a reduction of energy level.

Barrier attenuation is noise reduction due to a solid barrier located between a source and a receiver. A barrier such as a wall interrupts the direct sound propagation path. To achieve the maximum effectiveness, a barrier should be designed such that it interrupts the direct line of sight between the source and the receiver.

5.3.1.1.12 Sound Propagation Modeling

To estimate the highway noise impact, computer-based noise modeling programs are developed. These models include practical empirical engineering models, semiempirical mechanistic models, and numerical mechanistic models.

With the practical empirical engineering approach, methods rely on correlating actual field measurements of sound levels associated with a wide range of sound sources, terrain types (e.g., flat, hilly, grassy, paved), environmental conditions (e.g., wind speed, humidity, temperature), distances between sources and receivers, and other parameters. Vehicle speed, acceleration, deceleration, vehicle types, tail-pipe location and height, tire types, pavement types are also tested and correlated parameters. Regression equations and lookup tables are often deployed to relate sound levels to these parameters.

These practical empirical engineering models are computationally simple and easy to use. However, the simplifications associated with such models make them less accurate and sometimes difficult to apply to unique situations.

Empirical semi-mechanistic approaches are partially based on analytical solutions of wave equations. However, their reliance on empirical data is still a critical component in model development. Semi-mechanistic approaches often can model more complicated scenarios than pure empirical engineering models. The FHWA's Traffic Noise Modeling (TNM) program is an example of a semi-mechanistic modeling software.

The last modeling approach is the mechanistic numerical approach. While this type of model provides great flexibility in modeling sound propagation, its input requirements are substantial. Its sound level estimates are often less accurate than those of the other methods. The advantage of such mechanistic models is that their applications are less constrained by scenario conditions.

5.3.1.2 Noise Impact Analysis Practice

Noise impact analysis as related to highway projects follows the sequence of:

1. identifying potentially noise-sensitive sites (receivers),
2. determining actual receiver location coordinates (x, y, z) for these potential noise-sensitive sites,
3. coding geospatial relationship among roadways and receivers (x, y, z),
4. obtaining traffic data (volume, class, and speed) from the Design Traffic report covering the current-year, opening-year, and design-year traffic data,
5. entering the above data into the FHWA TNM model and run the model, and
6. interpreting TNM outputs.

If a projected noise level approaches 67 dBA (the common interpretation of approaching 67 dBA is when the actual numerical number exceeds 65 dBA), then appropriate noise abatement measures, including the feasibility of a noise wall, are analyzed.

For any noise wall modeling, a potential wall's spatial location in the same georeferenced system (x, y, z) used for roadway and receivers should be used. A noise wall's horizontal alignment typically follows the right of way line. Its vertical alignment follows the topography along the right of way. Analysts should systemically model a wall by adjusting its height (z) sequentially until the noise level at receivers is below the 67 dBA (65 dBA) predicted by the TNM.

Whether a noise wall should be built or not depends on a host of additional conditions such as those listed below.

1. Cost – State DOTs have limitations on the maximum amount of money that can be invested for a given receiver. If the cost exceeds the limit, a wall is considered not feasible.
2. Wall height – State DOTs may have limitations on what the maximum height of a wall may be. In addition to increased cost, a too-tall wall is often considered undesirable by residents.

3. Desires of impacted residents – If an impacted resident does not want to have a wall, the state may not force the resident to have one.

4. Effectiveness of a wall – A wall must be truly effective in providing noise reduction for an impacted resident. If a wall with its dimensions and locations cannot provide adequate insertion loss

NOSIE ATTENUATION THROUGH:

- geometric attenuation,
- atmospheric absorption,
- terrain ground surface bouncing, and
- barrier refection and absorption.

(e.g., 5 dBA), then the wall is deemed ineffective and should not be built.

5. Other site conditions (e.g., constructability) and limitations preventing a wall's feasibility.

6. Antiquity consideration – State DOTs may also consider a wall's feasibility based on the rationale of "who is there first." If a highway was there before a receiver site, the site might not qualify for a noise wall.

Highway noise impact is very controversial. Impacted citizens are often very emotional. Project engineers, project managers, and noise analysts must pay close attention to this issue. Alternative measures such as planting evergreen bushes and trees (no noise abatement capability) to soften the visual impact of a highway are often offered to sites that do not qualify for noise abatement walls.

5.3.2 Air Quality

The U.S. EPA under the authority of the Clean Air Act Amendment has established the National Ambient Air Quality Standards (NAAQS). To achieve and maintain the NAAQS for air quality non-attainment areas, state environmental agencies develop and implement State Implementation Plans (SIP). Additionally, transportation programs are required to pass the Transportation Conformity determination. For other non-classifiable areas, no actions as related to air quality are needed on a program level.

For any individual transportation project, a transportation agency is required to address impacts from carbon monoxide (CO) in all areas and particulate matter (PM) if the project is within a PM non-attainment area.

For CO impact analysis, CO concentration estimation through modeling covering human activity areas (e.g., sidewalk next to an intersection) is performed. The modeled CO concentration is compared to the NAAQS 1-hour 35 ppm concentration standard. If the modeled CO concentration exceeds the 35-ppm level, alternative design and traffic control will be needed to remedy the impact.

The PM project-level impact analysis is only applicable to highway projects within a PM non-attainment area. The analysis covers a qualitative evaluation of PM emissions from construction activities and how to control dust during construction.

5.3.2.1 Basic Terminology and Concepts

5.3.2.1.1 Concentration

A substance's concentration (C) in the air is the amount of the substance per unit of air.

$$C = \frac{\text{amount of substance measured in volume}}{\text{volume of total air}}$$

Parts per million, known as ppm, is one of the most commonly used measurement units in air quality analysis. PPM may have a unit of volume/volume or mole/mole. For example, 6 ppm means 6 volumes of a substance in 1 million volumes of total air or 6 moles of a substance in 1 million moles of total air. Occasionally, units such as g/m³ or µg/m³ are used. Given that different compounds have different molecular weighs, mass/volume unit specification requires additional conditions such as air pressure and temperature. For the above reason, the mass volume unit is used neither often nor widely.

5.3.2.1.2 Flux

The flux of a substance is the transport of the substance per cross-section area per unit time (Figure 5.6).

$$\text{flux} = \frac{\text{amount of substance passing through cross-section}}{\text{cross section area } X \text{ time}}$$

5.3.2.2 Dispersion Mechanism

Dispersion analysis or dispersion modeling is the study of the geospatial distribution of air pollutants as time elapses.

Air pollutant dispersion mechanisms include advection, diffusion, chemical reaction, wet and dry deposition, gravitational settling, and radioactive decay.

5.3.2.2.1 Advection

Advection refers to the passive transport of a substance from one place to another by moving air. For example, a particle is brought from location A to location B by the wind. Here, the particle's movement relies on the wind as its carrying medium.

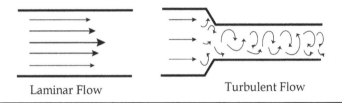

Laminar Flow   Turbulent Flow

Figure 5.6   Turbulent flow and laminar flow illustration.

5.3.2.2.2 Diffusion

Diffusion can be categorized into molecular diffusion and turbulent diffusion. Molecular diffusion is the process where a substance moves from one place to another through its own random perpetual motion of molecules. Fick's law is often used to quantify and analyze such diffusion.

Turbulent diffusion refers to mixing movements through eddy flows as defined in fluid dynamics and characterized by chaotic changes in pressure and flow velocity along a flow path. Turbulent diffusion does not exist in a laminar flow regime where the medium (e.g., air) flows in parallel layers without interfering with each other (see Figure 5.6 for illustration).

5.3.2.2.3 Chemical Reaction

The chemical reaction is the elimination of an existing substance and the production of new ones. VOCs react with NO_x under sunlight conditions and produce ozone (O_3). In this case, the disappearance of VOCs and NO_x is due to the complicated chemical reactions and through these chemical reactions, a new substance O_3 is produced.

5.3.2.2.4 Wet and Dry Deposition

Wet and dry deposition refers to acid rain, fog, snow, or particles where chemicals are either dissolved into or absorbed onto. Wet and dry depositions can be combinations of both chemical reactions and physical aggregation. For example, sulfur dioxide gas can be dissolved into rainwater and form sulfurous acid ($SO_2 + H_2O \rightleftarrows H_2SO_3$). Sulfurous acid can then be absorbed physically by dust particles.

5.3.2.2.5 Gravitational Settling

Gravitational settling is the process of substance separation due to gravitational force. Gravitational settling is associated with large PM such as dust and is strictly a physical process.

5.3.2.2.6 Radioactive Decay

Radioactive decay is a radioactive material's own destruction based on its half-life property. For example, the gaseous radon (Rn) alpha decay generates polonium (Po).

$$\substack{22\\86}\text{Rn} \rightarrow \substack{4\\2}\text{H} + \substack{216\\84}\text{Po}$$

For CO modeling, only advection and diffusion need to be considered. CO is a relatively stable compound, and its chemical reactions with other compounds in the air are slow. CO is also water-insoluble and not radioactive. It behaves very similarly to other air components such as nitrogen gas and oxygen gas.

5.3.2.3 Gaussian Dispersion Modeling

There are many modeling approaches for characterizing air pollutant dispersion. Gaussian dispersion modeling, with its straightforward single computation procedure, has been used in transportation air quality modeling software such as CALINE3 and CAL3QHC.

Air pollutant concentration $c(x, y, z)$ is affected by source strength (e.g., number of cars and trucks), source height (e.g., vehicle exhaust pipe height), downward wind speed, vertical mixing, and crosswind mixing characteristics.

$$c(x,y,z) = \frac{Q}{2\pi\sigma_y\sigma_z v} \exp\left\{\frac{y^2}{2\sigma_y^2}\right\}\left\{\exp\left\{\frac{-(z-h)^2}{2\sigma_z^2}\right\} + \exp\left\{\frac{-(z+h)^2}{2\sigma_z^2}\right\}\right\}$$

where

$\quad c(x, y, z)$: pollutant concentration at a given geospatial location of (x, y, z) with
$\qquad x$ along the downward wind direction, y the crosswind direction, and z in
$\qquad$ the vertical direction
$\quad Q$: pollutant source strength (emission rate from vehicles)
$\quad h$: pollutant source height (height of tailpipes)
$\quad v$: downward wind speed (wind blows toward a receiver, x-direction)
$\quad \sigma_y$ and σ_z: dispersion standard deviations from crosswind and vertical pollutant
$\qquad$ mixing

5.3.2.4 Practices in Transportation Air Quality Project-Level Modeling

For project-level CO analysis, the approach is to model the worst-case scenario. The basis of modeling the worst-case scenario is that if the worst case does not cause any violation, no other cases could cause air quality violation.

The worst-case scenario refers to the least pollutant dispersion condition. The worst-case condition consists of (a) downward wind speed is set at 1 foot per second where advection is minimal, (b) crosswind is set at zero, and (c) vertical mixing assumes the most stable meteorological class of A.

These assumptions result in the least pollutant dispersion and produce the highest pollutant concentration at a receiver site.

For roadway condition data, traffic volume is set as peak hour maximum with both mobile and idling emissions (each vehicle is an emission source). Total impacts to a receiver are the summation of each vehicle traveling along a roadway and within an intersection from all approaches.

To estimate CO concentration at a sidewalk (receiver), the sidewalk is purposely set so that it situated downwind (wind blows pollutants toward the site), thus experiencing the worst meteorological condition and worst traffic condition

CO CONCENTRATION CONTOUR

A carbon monoxide concentration contour on a map represents locations with similar CO concentration. A series of contour lines provides a concentration gradient.

as discussed above. If the modeled CO concentration is lower than 35 ppm, then the project has no air quality impact and no further action is needed. If the modeled CO concentration is higher than 35 ppm, then additional traffic data and local metrological data may be used for reassessment. If using local data still causes NAAQS CO standard violation, the roadway alignment or geometry shall be modified until air quality standard is met.

Highway air quality modeling software can also generate CO concentration contour lines. These contour lines can be used to determine whether a human activity area is impacted.

5.3.2.5 Greenhouse Gas Emission Estimation

In the United States, highway usage of diesel and gasoline was over 178 billion gallons (2020) and the trend is not coming down per the FHWA's Highway Statistics report. Highway transportation consumes the largest percentage of liquid petroleum products. Ultimately, the vast majority of this petroleum is converted into CO_2 through combustion or other eventual oxidation means. The best approach in computing CO_2 emission is through diesel and gasoline fuel data.

The greenhouse gas effect is a regional, national, and international phenomenon. Project-level greenhouse gas emission analysis vs. network-wide, region-wide, and national should be carefully considered. However, currently, there is no legal requirement to perform greenhouse gas analysis at any level of geography.

5.3.3 Contamination Assessment

The analysis and evaluation of the presence of hazardous material contained in soil, surface water, or groundwater within the right of way or proposed right of way of a roadway is called contamination assessment. Hazardous materials within the right of way or adjacent to a roadway may include gasoline, diesel, solvents, other petroleum products, lead, and other industrial pollutants.

The Federal Comprehensive Environmental Response, Compensation, and Liability Act of 1980 as amended provides a superfund for the cleanup of abandoned and uncontrolled hazardous contaminated sites as well as for emergency spills and releases. The Act requires early coordination with the U.S. EPA for proper cleanup and defines liability and liability transferability. The U.S. EPA cleans contaminated sites through various mechanisms, including orders, consent decrees, and settlements with responsible parties.

Contamination remediation is costly. Potential contamination effects should be evaluated during the right of way purchasing decision-making process.

5.3.3.1 Contamination Assessment Objective

The goal of contamination assessment is to protect the public from harmful hazardous material and ensure that a roadway project will not contaminate its surrounding environment. State DOTs achieve this goal by identifying potential hazardous material issue sites to avoid them first, planing appropriate remediation actions for unavoidable locations (e.g., removal of contaminated soil, treat gasoline contaminated groundwater...), and seeking appropriate cost reimbursement incurred for cleanup work.

5.3.3.2 Contamination Assessment Processes

Contamination assessment relies on a two-step process to gain a comprehensive understanding of the hazardous material issue. The first step is to check registered sites with the EPA and states' environmental regulatory agencies for background information.

Attention should be paid to gasoline stations with underground storage operations; agricultural sites where petroleum, pesticides, fungicides, and fertilizers are stored; dry cleaning shops where organic solvents are used; and salvage and auto junkyards where gasoline, oil, solvents, and organic solutions may be involved.

These sites listed above typically require EPA and state environmental regulatory agency permits to operate. Known spills, leaks, and other incidental releases of hazardous material are required to be reported to regulatory agencies. Also, regulatory agencies may have information as related to both past and present issues, including remediation or closure actions. Regulatory agency inspection reports, enforcement notices, and other publicly available information are highly useful.

The second step in assessing potential contamination impact is field review. Field review involves visiting actual sites, conversing with owners or operators, and interviewing others. After gathering and reviewing all the administrative background information, specialists perform a field examination on relevant parcels. Special attention should be paid to properties that are planned to be acquired. Specialists may also conduct interviews with residents and local government officials to ensure a comprehensive understating of any relevant issue.

5.3.3.3 Basic Remediation

Removal of contaminated soil is the most often encountered action during remediation work. Before any contaminated soil can be removed and disposed of, it is required to be treated. In addition to soil, groundwater contaminated by fuel and other organic solvents also requires remediation. Construction often involves dewatering, which disturbs groundwater flow patterns. By law, dewatering is considered an action causing further spreads of contaminants. Consequently, a party conducting the dewatering action is responsible for the cleanup. Usually, contaminated water is pumped out of the ground into a settling tank. From there, the water is filtered through active carbon to remove contaminants. Only when the treated water meets regulatory standards, it can then be discharged.

5.3.4 Utility Assessment

Utilities in the context of highway projects are infrastructures that provide services other than vehicle or pedestrian travel to the public and reside within a roadway's right of way. Examples include water mains, storm and wastewater sewer lines, power lines, natural gas lines, telecommunication lines, and others. Utility infrastructures can be owned and operated by either public agencies or private businesses.

Both Federal and State governments recognize that it is in the public's interest for utility facilities to be accommodated within the public highway right of way (provided that they do not interfere with highway functions). There are strict regulations related to the duties and responsibilities of both highway agencies and utility owners and providers. Government agencies such as State transportation departments and the U.S. DOT have very specific statutory rules and regulations to follow in dealing with utility issues (Figure 5.7).

5.3.4.1 Utility Assessment Objective

During the EIE phase, it is critical to identify utilities that existed within the project limit. The objectives of a utility impact assessment include:

- ensuring that impacts to utility facilities are known,
- offering adequate time for individual utility owner or operator to plan on needed actions,
- addressing cost associated with utility removal, adjustment, and other actions, and
- determining the exact geospatial locations to guide construction work.

Figure 5.7 **Examples of aboveground (power and other lines) and underground (water main) utilities within a highway right of way.**

5.3.4.2 Assessing Procedure

While the public highway right of way does offer opportunities to accommodate different utilities, a stipulation as required by law is that cost associated with adjustment, removal, and relocation of any utility as directed by state highway agencies be borne by utility owners or providers except in the case of Federal-aid Interstate highway right of way.

Utility relocation cost within the Federal-aid Interstate highway right of way may be paid by a state transportation department under certain conditions. One of these conditions is that Federal-participation in the project is at least 90%.

When a utility owner and provider is a local government (e.g., county, city, township), obligations of the local government are no different than any private business with regard to compliance with related law and regulation. When a local government does not have the needed fund for its utility relocation, the state transportation department can sign a legal reimbursement agreement with the local government. The legal reimbursement agreement allows the state transportation department to proceed with the relocation. The local government will reimburse the State at a later specified time.

5.3.4.3 Utility Impact Assessment Process

1. Prepare project location maps and drawings, marking a project's starting and ending points. Aerial photo maps with the right of way lines drawn are desired.

2. Request information from utility providers serving the area where the roadway project is located. Information requested should cover both aboveground and underground, existing, and planned utilities. Additional information as related to property rights or usage rights – such as easements, should also be included.
3. Hold a separate meeting with each utility provider and provide clear information ensuring both sides are knowledgeable on the issue.
4. Prepare utility relocation cost estimates and communicate with relevant utility owner or operator.
5. Notify both design engineers and utility owners and operators on the conclusion of the assessment.

Utility accommodation and relocation are complex. Duties and responsibilities for all involved parties vary from state to state. If utility accommodation and relocation issues are not adequately addressed before construction starts, it may lead to severe problems for a project during construction. An unresolved utility issue could significantly impact both project costs and schedules.

5.4 Final Products and Deliverable

The effort for analyzing different environmental topics varies depending on the complexity of the issue involved. Documentation for such effort can range from a simple paragraph to a report with hundreds of pages. After all environmental topic areas are analyzed and documented, a summary environmental evaluation analysis report is prepared.

If a project is classified as Type II Categorical Exclusion (CE), the environmental impact summary report along with the Preliminary Engineering Report is sent to the FHWA for review and approval. Upon approval of the environmental document by the FHWA, the project is qualified for Federal funding, and the project can be moved to final engineering design and other phases.

If a project is classified as Environmental Assessment (EA), and if the environmental analysis concludes that environmental impact is significant, then the EA will be reclassified as (bump-up) an Environmental Impact Statement (EIS). The administrative procedure for the EIS process is then followed. Otherwise, if the EA analysis concludes that environmental impact is not significant, a Finding of No Significant Impact (FONSI) report is prepared for final Federal approval.

When a project is classified as EIS, the EIS administrative procedure is to be followed. First, a Draft EIS is prepared for public comment. Upon completion of public commenting, a Final EIS is prepared. Again, the public is notified for further review and comment. Upon the conclusion of the Final EIS, the U.S. DOT may issue a Record of Decision (ROD) for the final approval of the project.

The Federal approval of the summary environmental document is known as granting Location Design approval. All these individual environmental evaluation reports (e.g., highway noise, wildlife and habitat, wetland), environmental summary reports, and the Preliminary Engineering Report are part of the administrative record demonstrating Federal NEPA compliance.

Figures 5.8a is an example of a Federal Register Notice announcing the public availability of an EIS document. The U. S. EPA manages all Federal Agency's EIS

Federal Register / Vol. 78, No. 125 / Friday, June 28, 2013 / Notices   **38975**

CARB that the TRU amendments do not undermine California's protectiveness determination from its previous authorization request. Second, EPA agrees with CARB that California's TRU amendments do not undermine EPA's prior determination regarding consistency with section 202(a) of the Act. Third, EPA agrees with CARB that California's TRU amendments do not present any new issues which would affect the previous authorization for California's TRU ATCM regulations. Therefore, I confirm that CARB's TRU amendments are within the scope of EPA's authorization for California's TRU ATCM regulations.

My decision will affect not only persons in California, but also manufacturers outside the State who must comply with California's requirements in order to produce TRU systems for sale in California. For this reason, I determine and find that this is a final action of national applicability for purposes of section 307(b)(1) of the Act. Pursuant to section 307(b)(1) of the Act, judicial review of this final action may be sought only in the United States Court of Appeals for the District of Columbia Circuit. Petitions for review must be filed by August 27, 2013. Judicial review of this final action may not be obtained in subsequent enforcement proceedings, pursuant to section 307(b)(2) of the Act.

IV. Statutory and Executive Order Reviews

As with past authorization and waiver decisions, this action is not a rule as defined by Executive Order 12866. Therefore, it is exempt from review by the Office of Management and Budget as required for rules and regulations by Executive Order 12866.

In addition, this action is not a rule as defined in the Regulatory Flexibility Act, 5 U.S.C. 601(2). Therefore, EPA has not prepared a supporting regulatory flexibility analysis addressing the impact of this action on small business entities.

Further, the Congressional Review Act, 5 U.S.C. 801, *et seq.*, as added by the Small Business Regulatory Enforcement Fairness Act of 1996, does not apply because this action is not a rule for purposes of 5 U.S.C. 804(3).

Dated: June 19, 2013.

Gina McCarthy,
Assistant Administrator, Office of Air and Radiation.

[FR Doc. 2013–15437 Filed 6–27–13; 8:45 am]
BILLING CODE 6560-50-P

ENVIRONMENTAL PROTECTION AGENCY

[ER–FRL–9009–8]

Environmental Impacts Statements; Notice of Availability

Responsible Agency: Office of Federal Activities, General Information (202) 564–7146 or *http://www.epa.gov/ compliance/nepa/.*
Weekly receipt of Environmental Impact Statements
Filed 06/17/2013 Through 06/21/2013
Pursuant to 40 CFR 1506.9.

Notice

Section 309(a) of the Clean Air Act requires that EPA make public its comments on EISs issued by other Federal agencies. EPA's comment letters on EISs are available at: *http:// www.epa.gov/compliance/nepa/ eisdata.html.*
EIS No. 20130178, Draft EIS, USACE, FL, Port Everglades Harbor Navigation Improvements, Comment Period Ends: 08/13/2013, Contact: Terri Jordan-Sellers 904–232–1817.
EIS No. 20130179, Draft EIS, BLM, WY, Buffalo Field Office Planning Area Resource Management Plan, Comment Period Ends: 09/28/2013, Contact: Thomas Bills 307–684–1133.
EIS No. 20130180, Draft EIS, BLM, WAPA, 00, TransWest Express Transmission Project, Comment Period Ends: 09/25/2013, Contact: Sharon Knowlton 307–775–6124.
The U.S. Department of the Interior's Bureau of Land Management and the U.S. Department of Energy's Western Area Power Administration are joint lead agencies for the above project.
EIS No. 20130181, Final EIS, USAF, AK, Modernization and Enhancement of Ranges, Airspace and Training Areas in the Joint Pacific Alaska Range Complex in Alaska, Review Period Ends: 07/29/2013, Contact: Tania Bryan 907–552–2341.
EIS No. 20130182, Draft EIS, EPA, LA, Designation of the Atchafalaya River Bar Channel Ocean Dredged Material Disposal Site West, Pursuant to Section 102(c) of the Marine Protection, Research, and Sanctuaries Act of 1972, Comment Period Ends: 08/12/2013, Contact: Jessica Franks 214–665–8335.
EIS No. 20130183, Final Supplement, NRC, NY, Generic—License Renewal of Nuclear Plants, Supplement 38, Regarding Indian Point Nuclear Generating Unit Nos. 2 and 3, Review Period Ends: 07/29/2013, Contact: Lois James 301–415–3306.

EIS No. 20130184, Draft Supplement, FHWA, AK, Gravina Access Project, Comment Period Ends: 08/13/2013, Contact: Kris Riesenberg 907–465–7413 EIS No. 20130185, Draft Supplement, Caltrans, CA, San Diego Freeway (I–405) Improvement Project, Comment Period Ends: 08/12/2013, Contact: Smita Deshpande 949–724–2000.

Dated: June 25, 2013.

Cliff Rader,
Director, NEPA Compliance Division, Office of Federal Activities.

[FR Doc. 2013–15612 Filed 6–27–13; 8:45 am]
BILLING CODE 6560-50-P

EXPORT-IMPORT BANK OF THE UNITED STATES

Sunshine Act Meetings

ACTION: Notice of a partially open meeting of the Board of Directors of the Export-Import Bank of the United States.

TIME AND PLACE: Tuesday, July 9, 2013 at 9:30 a.m. The meeting will be held at Ex-Im Bank in Room 321, 811 Vermont Avenue NW., Washington, DC 20571.

OPEN AGENDA ITEMS: Item No. 1: 2013 Review of the Content Policy.

PUBLIC PARTICIPATION: The meeting will be open to public observation for Item No. 1 only.

FURTHER INFORMATION: Members of the public who wish to attend the meeting should call Joyce Stone, Office of the Secretariat, 811 Vermont Avenue NW., Washington, DC 20571 (202) 565–3336 by close of business Monday, July 8, 2013.

Cristopolis A. Dieguez,
Program Specialist, Office of the General Counsel.

[FR Doc. 2013–15702 Filed 6–26–13; 4:15 pm]
BILLING CODE 6690-01-P

FEDERAL COMMUNICATIONS COMMISSION

Sunshine Act Meetings

Open Commission Meeting

Thursday, June 27, 2013

The Federal Communications Commission will hold an Open Meeting on the subjects listed below on Thursday, June 27, 2013. The meeting is scheduled to commence at 10:30 a.m. in Room TW–C305, at 445 12th Street SW., Washington, DC.

Figure 5.8a   An example on EIS availability notice to the public via the Federal Register. (Courtesy of the Federal Register at www. federalregister.gov.)

announcements in the Federal Register. The Federal Register is a daily record of the Federal government published every business day by the National Archives and Records Administration (an independent U.S. Federal agency).

Figure 5.8b shows examples of DEIS and FEIS cover pages. A cover page contains the project title, lead agencies, cooperating agencies, various projects, and document information along with officials who authorized the actions.

Figure 5.8c shows an example of an EIS table of contents. The topic areas are extensive. However, depending on the specific topic, it can be either lengthy or brief.

5.5 Next Step

Once a highway or transit project receives the FHWA/FTA Location Design approval, the project is eligible for Federal funds for its final engineering design, right of way acquisition, and construction. The most common step following the Location Design approval is the final engineering, where a complete set of construction plans is developed.

5.6 Summary

EIE requires collaboration across a wide range of professionals. Unlike quantitative engineering computation and analysis, some of the environmental impact areas are often considered to be and viewed as subjective. Environmental impact evaluation results are also often controversial because people commenting on such analysis often have philosophical and belief-oriented observation biases. This phenomenon demands that environmental impact analysis is based on established legal and scientific procedures and processes.

Many state and federal resource agencies are involved in EIE and approval process. Often the process is long, and coordination is complicated. Congress recognizes the issue and through surface transportation legislation, including the FAST Act, has established different mechanisms to streamline the environmental impact analysis (the so-called streamlining NEPA), review, and permitting process to expedite transportation project development.

Engineers and program managers need to be familiar with relevant environmental issues, regulations, and laws. Project engineers and managers are challenged to balance all these competing and conflicting factors to achieve a viable solution to meet transportation needs and budget limitations.

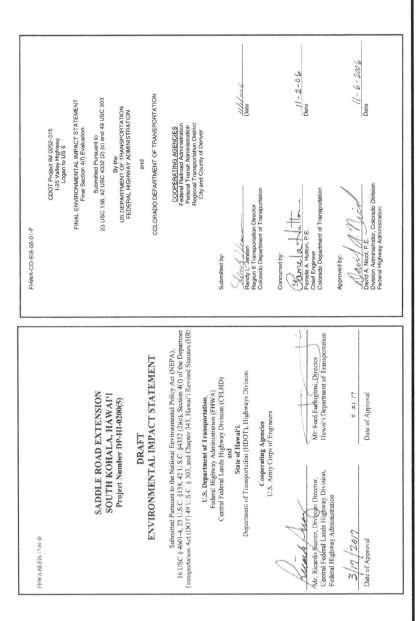

Figure 5.8b Example of draft EIS and final EIS cover pages. (Courtesy of 1) the Central Federal Lands Highway, the FHWA and HI DOT at www.flh.fhwa.dot.gov, and 2) the FHWA and CDOT at www.codot.gov.)

Valley Highway Logan To 6th Ave.
Environmental Impact Statement (EIS)

TABLE OF CONTENTS

	Page
EXECUTIVE SUMMARY	ES-1
1.0 PURPOSE AND NEED	**1-1**
1.1 Project Location and Purpose	1-1
1.2 Project History and Status	1-3
1.3 Project Needs and Objectives	1-5
1.4 Detailed Identification of the Project Needs	1-6
2.0 ALTERNATIVES	**2-1**
2.1 Public and Agency Involvement in Alternatives Development	2-1
2.2 Alternatives Development and Screening Process	2-4
2.3 Alternatives Considered and Eliminated	2-7
2.4 System Alternatives Considered in Detail	2-22
2.5 Future Redevelopment and Transportation Improvements Near I-25 and Broadway	2-63
2.6 Preferred Alternative	2-65
3.0 TRANSPORTATION ANALYSIS	**3-1**
3.1 Existing Roadway and Traffic Conditions	3-1
3.2 Compatibility with Transportation Plans and Programmed Projects	3-8
3.3 Future Travel Demand	3-9
3.4 Freeway and Street Safety	3-33
3.5 Transit / HOV Access	3-34
3.6 Pedestrian and Bicycle Facilities	3-35
3.7 Freight and Rail Operations	3-37
4.0 ENVIRONMENTAL CONSEQUENCES	**4-1**
4.1 Socio-Economics and Community	4-1
4.2 Right-of-Way and Displacements	4-2-1
4.3 Parks and Recreation	4-3-1
4.4 Aesthetics and Urban Design	4-4-1
4.5 Air Quality	4-5-1
4.6 Noise and Vibration	4-6-1
4.7 Historic Preservation	4-7-1
4.8 Paleontology	4-8-1
4.9 Water Resources	4-9-1
4.10 Floodplains	4-10-1
4.11 Wetlands, Waters of the U.S. and Open Water	4-11-1
4.12 Vegetation and Wildlife	4-12-1
4.13 Hazardous Waste	4-13-1

Valley Highway Logan To 6th Ave.
Environmental Impact Statement (EIS)

TABLE OF CONTENTS (Continued)

	Page
4.14 Soils and Geology	4-14-1
4.15 Energy	4-15-1
4.16 Short-term Uses and Long-term Productivity	4-16-1
4.17 Irreversible and Irretrievable Commitments of Resources	4-17-1
4.18 Construction Impacts	4-18-1
4.19 Permits Required	4-19-1
4.20 Cumulative Impacts	4-20-1
4.21 Summary of Environmental Consequences, Mitigation Measures and Monitoring Commitments	4-20-1
5.0 SECTION 4(f) EVALUATION	**5-1**
5.1 Parks	5-6
5.2 Coordination	5-21
5.3 Section 4(f) Finding	5-21
6.0 PUBLIC INVOLVEMENT	**6-1**
6.1 Objectives	6-1
6.2 Elements of Program	6-1
6.3 Agency Input	6-6
6.4 Public Input	6-11
6.5 Special Outreach to Low-Income and Minority Populations	6-20
6.6 Release of Draft EIS	6-25
6.7 Coordination Subsequent to Release of Final EIS	6-26
7.0 PHASED PROJECT IMPLEMENTATION	**7-1**
7.1 Phased Implementation Requirements	7-1
7.2 Identification of Logical Project Phases and Priorities	7-4
7.3 Detailed Discussion of Project Phases	7-7
7.4 Further Coordination and Decision Making after Final EIS	7-43
8.0 LIST OF PREPARERS	**8-1**
9.0 AVAILABILITY OF TECHNICAL REPORTS	**9-1**
10.0 REFERENCES	**10-1**
11.0 INDEX	**11-1**

APPENDIX A   AGENCY COORDINATION
APPENDIX B   PUBLIC COORDINATION
APPENDIX C   WETLAND FINDING

Figure 5.8c   Table of contents example of a final EIS. (Courtesy of the FHWA and CODOT at https://www.codot.gov/library.)

5.7 Discussion

1. What does "environmental streamlining" mean to you?
2. It routinely takes more than 24 months to complete an Environmental Impact Statement (EIS). During this period, many alternatives are studied on paper with minimum fieldwork. Given what you know, can you list the pros and cons of this lengthy process?
3. What does community cohesion mean to you?

5.8 Exercises

1. Environmental impact evaluation, as related to transportation project development, is an effort to comply with the National Environmental Policy Act. In other words, environmental impact evaluation is required by law. Is the above statement correct? Yes or No.
2. Environmental impact evaluation is often divided into three main areas knowns as (a) human, social, and cultural environment, (b) natural environment, and (c) physical environment. True or False.
3. The consideration of land use during a highway environmental impact evaluation is to ensure that highway development is compatible with the objective of locally developed and implemented land-use plans. True or False.
4. Highways should avoid splitting neighborhoods. Engineers need to pay close attention to communities and consider providing appropriate mechanisms such as overpasses and underpasses to ensure residents can cross a highway without any major inconvenience. True or False.
5. Most poured concrete for retaining walls, bridges, and other aboveground structures has its natural grayish color. Residents surrounding such facilities often desire alternative colors. Shall design engineers and project managers take such requests seriously? Yes or No.
6. For highway projects, relocation is often inevitable. Design engineers and program managers should minimize relocation. Relocation not only is costly but also causes anxiety to residents. True or False.
7. Minimizing and avoiding impacts to historic properties is required by law (Section 106 of the National Historic Preservation Act). True or False.
8. Section 106 of the National Historic Preservation Act requires public parks to be protected. True or False.
9. Section 4(f) of the U.S. DOT Act requires the U.S. Department of Transportation to protect public parks and recreational areas. True or False.
10. Historically significant public parks are protected by both Section 106 of the National Historic Preservation Act and Section 4(f) of the U.S. DOT Act. True or False.
11. The U.S. Fish and Wildlife Service is responsible for wetland impact evaluation and permitting. True or False.

12. The U.S. Army Corps of Engineers is responsible for wetland impact evaluation and permitting. True or False.

13. Wetland provides a unique ecosystem where a wide range of biological activities occur, ranging from water purification to acting as a habitat for a variety of plants and animals. True or False.

14. Define wetland mitigation bank.

15. Define wetland jurisdictional boundaries.

16. Define wetland mitigation ratio.

17. Which Federal agency is responsible for wildlife and habitat evaluation and analysis?
    - U.S. Fish and Wildlife Services
    - U.S. Department of Health and Human Services
    - U.S. State Department

18. An "action area" used in the context of wildlife and habitat analysis refers to the direct construction zone within a project's right of way. In other words, an "action area" cannot be larger than the right of way zone. True or False.

19. When a floodplain area is filled, its capacity to store water is reduced. Thus, the potential of flooding (water escaping from the floodplain) other areas during a storm is increased. Therefore, there are laws and regulations to prevent the encroachment of floodplains by transportation projects. True or False.

20. Stormwater runoff from highways is typically retained in stormwater retention or detention ponds for some minimum time before it can be discharged to receiving water. Stormwater retention and detention ponds serve dual roles in preventing flooding and improving water quality. True or False

21. Which agency is responsible for performing highway noise impact analysis for a highway-widening project?
    - State highway agency
    - State environmental agency
    - U.S. EPA

22. What is the highway noise abatement criteria level for a residential home under Federal regulation?

23. Sound "insertion loss" associated with highway noise abatement shows the effectiveness of an abatement measure. The bigger the "insertion loss," the more effective the abatement measure. True or False.

24. Noise measuring equipment for highway field noise monitoring should be equipped with the A-scale weighting function. True or False.

25. Explain the highway noise attenuation mechanism.

26. Increasing the distance between a highway and a home is very effective in reducing highway noise impact. This phenomenon is a result of the so-called "geometric attenuation" effect. True or False.

27. What is the dispersion mechanism for air pollutants?

28. What does NAAQS stand for?

29. The NAAQS standards for carbon monoxide (CO) are 35 ppm for the hourly standard and 9 ppm for the 8-hour period, respectively. What is the logical

explanation that the 1-hour standard level is significantly higher than the 8-hour concentration standard?

30. To deal with stormwater runoff from a highway, stormwater retention and detention ponds are constructed. What are the two key reasons for stormwater retention and detention ponds?

31. List the types of utility infrastructures within a highway right of way you can see. List all others you think might be within a highway right of way but can't be seen by simply looking around.

32. Thirty-six (36) motorcycles traveling along a stretch of highway generate 67 dBA level of noise at a resident's backyard. Now, 72 motorcycles are traveling on the same highway segment. The new noise level is expected to be $2 \times 67 = 134$ dBA. Is the above 134 dBA correct?

33. A highway project impacts 0.8 acres of wetland. The mitigation ratio agreed between the transportation department and the regulatory resource agency is 1.2. How many acres of new wetland does the Transporation department need to compensate?

34. The public availability announcement of any EIS can always be found from the Federal Register. True or False.

35. During an environmental impact evaluation process, it is highly possible that while trying to avoid impact to one resource, the solution may lead to a bigger or more severe impact on another resource. And it is unlikely that there is such an alternative where all environmental impacts on every subject area are the least. Situations like what described above require project managers, engineers, and other specialists to work together to derive a balanced alternative. True or False.

Bibliography

Environmental Analysis & Review, FTA, https://www.transit.dot.gov/regulations-and-guidance/environmental-programs/environmental-analysis-review.

Environmental Review Toolkit, FHWA, https://www.environment.fhwa.dot.gov/.

Environmental Review-Stormwater, Floodplain, and Coastal Zone Management, FHWA, https://www.environment.fhwa.dot.gov/env_topics/water.aspx.

Environmental Review-Wildlife and Habitat, FHWA, https://www.environment.fhwa.dot.gov/env_topics/wildlife.aspx.

Highway Traffic Noise: Analysis and Abatement Guidance, FHWA, https://www.fhwa.dot.gov/environment/noise/regulations_and_guidance/analysis_and_abatement_guidance/revguidance.pdf.

Overview of NEPA as Applied to Transportation Projects, FHWA, https://www.fhwa.dot.gov/federal-aidessentials/catmod.cfm?category=environm.

Transportation Conformity, FHWA, https://www.fhwa.dot.gov/ENVIRonment/air_quality/conformity/.

Chapter 6

Final Engineering Design

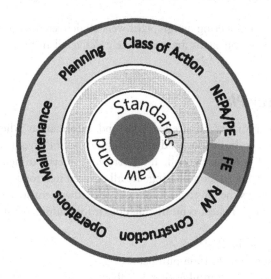

During the environmental impact evaluation (EIE) and preliminary engineering (PE) phase, a location design integrating both horizontal and vertical alignments for a roadway is developed. After the project receives the lead Federal agency approval for its environmental impact evaluation, the project is qualified for Federal funding covering all subsequent phases, including the immediate next final engineering design. The final engineering design continues the preliminary engineering design and concludes when all design work is complete.

6.1 Final Engineering Design Deliverables

Upon completion of the final engineering design, several products are produced and delivered. These products include but are not limited to those listed below.

■ Roadway Plan
■ Structure Plan
■ Signing and Pavement Marking Plan
■ Street Lighting Plan
■ Intelligent Transportation System (ITS) Plan
■ Traffic Signalization and Control Plan
■ Landscaping Plan
■ Proposed Right of Way Plan

Historically, final engineering design plans were all delivered with ink paper prints. Nowadays, construction plans are more likely delivered in both ink paper prints and electronic files.

6.1.1 Roadway Plan

A roadway plan provides all needed data and information where construction personnel can use to build the road accordingly. Specific layout, data, and information, such as alignment, cut, fill, and borrow, geometric dimensions, pavement information, and construction phasing, are all covered in a roadway plan.

Figure 6.1a shows the cover sheet for a set of roadway construction plan.

A design plan cover sheet typically includes information, as listed below.

■ Ownership of the Project
■ State project identification number
■ Federal-Aid project number (if applicable)
■ Project location county or counties
■ Brief project description
■ Project limits
■ Location map with project route and limits identified and illustrated
■ Index of different design sheet component sheets
■ Design traffic data
■ Design engineer signature and seal

In the example illustrated in Figure 6.1a, the cover sheet provides the following information:

■ Ownership – Ohio Department of Transportation (DOT)
■ State project number – WAY-30-9.11

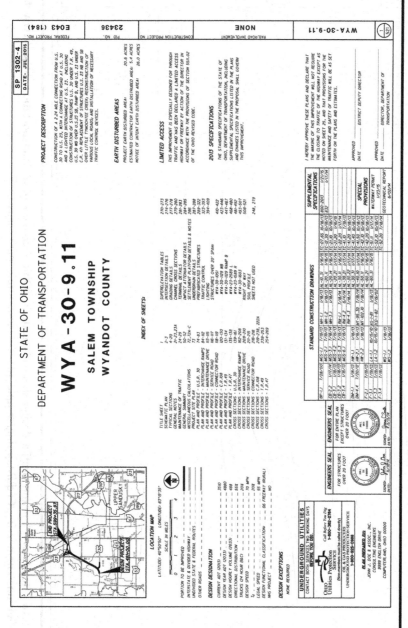

Figure 6.1a   Roadway design plan cover sheet illustration. (Ohio Department of Transportation, http://www.dot.state.oh.us/ Divisions/Engineering/CaddMapping/sample_plans/sample_plans/Pages/sample_plans.aspx.)

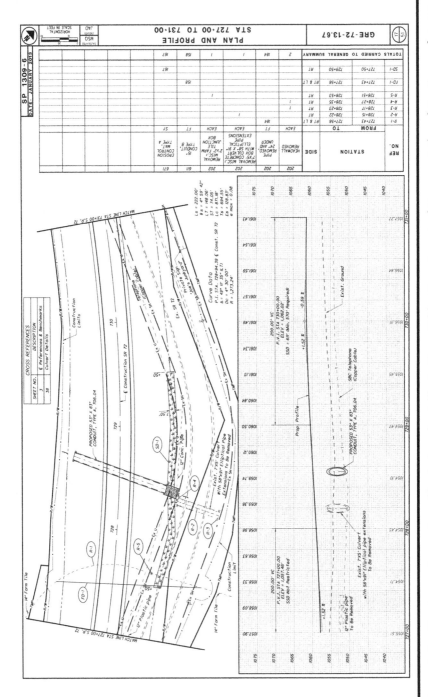

Figure 6.1b Roadway plan and profile design sheet illustration. (Ohio Department of Transportation, http://www.dot.state. oh.us/Divisions/Engineering/CaddMapping/sample_plans/Pages/sample_plans.aspx.)

- Federal-aid project number (if applicable) – E043 (184)
- Project location county or counties of the state – Wyandot County, Salem Township
- Project description – Construction of a 2.24-mile connection road and others
- Project limit – from US Highway 30 to US Highway 23
- Location map – Top right corner of the sheet with specific location highlighted
- Index of design sheets: Various components covered by sheet 1 to sheet 521
- Design traffic data: Current 2013 traffic data and projected 2033 traffic data
- Design engineer signature and seal – Engineer of records for both the roadway design and structure design signatures and seals. The signature and seal of the structures engineer are due to the fact that there is a structural component for this project.

Figure 6.1b illustrates a plan and profile sheet in a typical roadway design plan. The top part of the plan and profile sheet always shows the plan view where a roadway's layout is shown in relation to other referencing marks such as the right of way lines, buildings, connected roadways, and all surficial geometric features. The bottom part is the corresponding roadway profile. The profile section shows elevation information for both the existing ground and the proposed profile, underground cross drain locations, and other information such as various utilities crossing a highway.

The plan and profile sheet No. 1309-7(a) shown in Figure 6.1b covers a horizontal distance of 400 feet starting from Station 727+00 to Station 731+00 for SR72.

The plan view shows that the 400-feet segment is on a new alignment. And the entire segment is on a horizontal curve with its P.I. located at Station 728+94. The curve has a radius of 1,273.24 feet.

The original alignment is to be abandoned with a cross drain structure (7′ × 5′ box culvert) to be removed and a new large 53″ × 83″ conduit to be constructed for stormwater drainage. Construction activity shall stay within the construction limit as shown. Construction limits are inside and away from the right of way boundary.

The profile view shows that the new SR72 has a much higher elevation than the existing ground. At Station 726+00, the to-be-constructed new SR72 profile is 2.39 feet higher than the existing ground (1,057.90 − 1,055.51). At Station 731+00, the existing ground has an elevation of 1,057.27. The new alignment at the same location has an elevation of 1,061.43, which is 4.16 feet higher.

Two vertical curves are used for the vertical alignment. The first one is centered at Station 727+00.00 and extends to Station 728+00. The second one is centered at station 737+00.00, extending 100 feet in either direction between Station 729+00 and Station 731+00.

This plan and profile sheet also contains quantity information, as summarized in the table provided. The quantitative table provides specific information on earth to be moved, the quantity of standard construction pay items needed, and other materials needed.

Figure 6.1c illustrates a cross-section sheet associated with a roadway design plan. Cross-section sheets provide (a) detailed elevation data for a road, relating the roadway's proposed profile to existing grounds or existing roadways, (b) information on how to tie down construction limits, cut and fill pattern and extent, and (c) location and types of utilities buried longitudinally within the right of way.

Depending on the complexity of the design, cross-section sheets are typically provided at a minimum of every full station (100 feet). For urban roadways with complex design, cross-section sheets are often provided at intervals of 25 feet or 50 feet.

The three cross sections shown in Figure 6.1c provide the information listed below.

■ The three cross sections are located at Stations 784+00, 784+50, and 7855+00.
■ Elevations between cross sections can be linearly extrapolated.
■ The roadway to be constructed is on a new alignment where existing ground elevation is much lower than the roadway profile.
■ The three cross sections show a curb gutter urban roadway design possessing three lanes in each direction, a raised (flat) median, a 10-feet sidewalk next to the curb in both directions.
■ The sidewalk has a 2% cross slope, draining stormwater toward the curb. The three traveled lanes and the shoulder have a cross slope of 2%, draining stormwater toward the outside lane curb.
■ While fill is needed to raise the profile for the traveled way, a stormwater drainage ditch is to be constructed by cutting down the existing earth.
■ Utilities are present at the locations identified. These utilities are gas line, fiber optic cable, and electrical cable. Be aware that the locations for these utilities are not guaranteed to be precise. Construction in the vicinity of these utilities should be carried out with great diligence to avoid utility damage.
■ In addition to the existing R/W marking, there is also an existing easement beyond the right of way line. However, construction activities shall not occur beyond the ditch.

6.1.2 Structure Plan

If a project has a structural component (e.g., a bridge), a set of structure plan is prepared in order to provide all necessary information for both the construction of the substructure and superstructure.

THE FEDERAL AGENCY ISSUES BRIDGE PERMITS

The U.S. Coast Guard in the Department of Homeland Security issues permits for bridges over navigable waterways of the United States

---

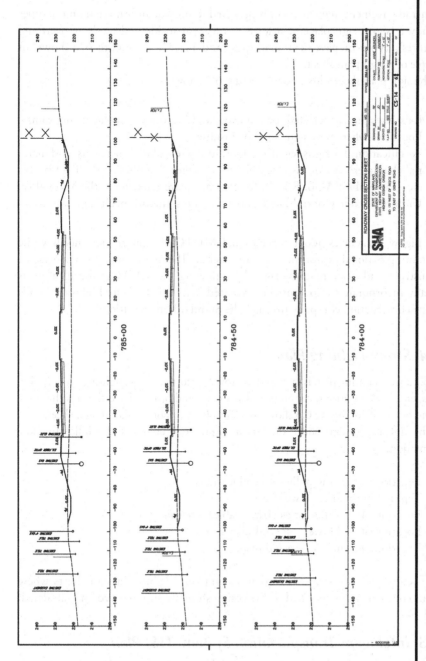

Figure 6.1c **Illustration of one of many cross-sectional sheets. (SHA, Maryland Open Bid, https://emaryland.buyspeed.com/bso/ external/bidDetail.sdo?bidId=MDJ0231027399&parentUrl=a.)**

6.1.3 Sign and Pavement Marking Plan

A sign and pavement marking plan is prepared to provide information on the types of roadway signs, sign placement locations (e.g., roadside and roadway overhung structures), types of pavement markings, and placement. Figure 6.2 shows an example of a sign plan sheet.

The sign plan sheet illustrated in Figure 6.2 shows:

■ Existing signs, which shall be removed, and locations of such existing signs
■ Locations where new signs shall be installed
■ Specifications for signs regarding font, font size, color, dimensions, and material characteristics are also specified in the notes "R3H8BA-36", "R3-9B-36", etc. The R3H8BA-36, R3-9B-36, etc., notes are referencing the Manual on Uniform Traffic Control Devices (MUTCD) standard specifications.

State highway agencies adopt the national MUTCD to meet local needs while maintaining national uniformity and consistency. The requirement for a contractor to construct and place roadway signs by following the MUTCD specification is typically incorporated into contracts executed between the State highway agency and the construction company through the Standard Specification.

6.1.4 Street Lighting Plan

If applicable, a lighting plan is prepared to provide information on lighting types (poles, lamps, etc.), pole locations, conduits and switches, and other related fixtures that may need to be installed. Figure 6.3 shows an example of a lighting plan design sheet.

The lighting design sheet shown in Figure 6.3 contains the following key information.

1. Location of lighting polls – seven locations
2. Types of luminaires – two types
3. Location of conduits for electrical cables – several places
4. Location of pull boxes – several places
5. Location of trenches – several places

There is always a summary sheet that accompanies lighting design sheets where various components as specified in the design sheet are summarized and tabulated.

6.1.5 Intelligent Transportation System (ITS) Plan

If applicable, a set of ITS plans is prepared to provide information on the type and placement location for traffic monitoring and information display.

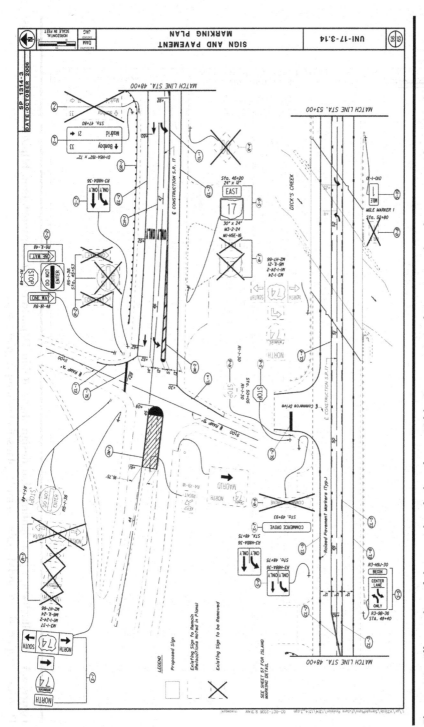

Figure 6.2 Illustration of signage plan design sheet. (Ohio Department of Transportation, http://www.dot.state.oh.us/Divisions/ Engineering/CaddMapping/sample_plans/Pages/sample_plans.aspx.)

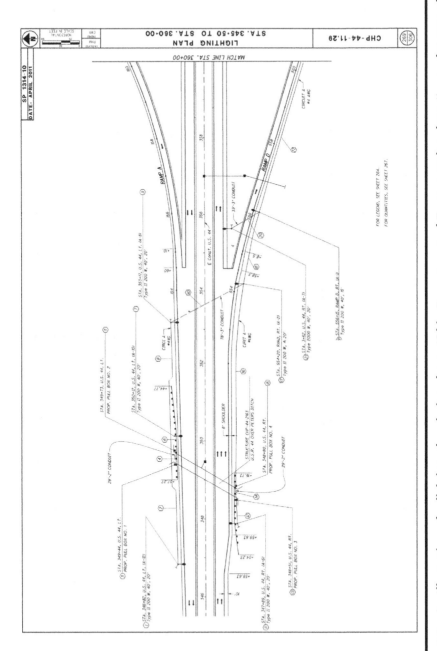

Figure 6.3 An Illustration of a lighting plan design sheet. (Ohio Department of Transportation, http://www.dot.state.oh.us/Divisions/Engineering/CaddMapping/sample_plans/Pages/sample_plans.aspx.)

6.1.6 Traffic Signalization and Control Plan

A traffic signalization and control plan is prepared to provide all needed information on traffic signal equipment installations associated with signalized intersections and other traffic control operations.

6.1.7 Landscaping Plan

A landscape plan is prepared to provide needed information on landscaping as related to types of greeneries, artifacts, and their locations. Figure 6.4 shows an example of a landscaping design plan sheet.

The landscaping sheet illustrated in Figure 6.4 covers the roadway segment from Stations 128+00 to 142+00, a distance of 1,400 feet. It has four separate wildflower areas covering 1.46, 0.94, 1.38, and 1.57 acres, respectively. These areas need to be seeded and mulched.

Also, five different trees with different sizes are specified for various locations. The specific quantity is summarized in the summary table. The placement of such trees shall follow the plan locations. Setback distances for these trees are specified in the design plan and should be followed.

Standard Specifications typically specify the need to maintain trees and other plants planted for a defined period. It is the contractor's responsibility to ensure all planted trees, shrubs, and others to survive and grow according to the standard specification.

6.1.8 Proposed Right of Way Plan

A proposed right of way plan is prepared in order to provide property boundary information covering both existing and newly proposed right of way. Right of way agents rely on the right of way map to acquire additional land.

6.1.9 Environmental Permits and Other Permits

All environmental permits, except the construction National Pollutant Discharge Elimination System (NPDES) permit (construction stormwater permits), should be secured from the U.S. EPA (or the EPA delegated State environmental regulatory agency) and the U.S. Army COE upon the completion of the final engineering design. A Coast Guard permit will need to be secured for any bridge over navigable waterways of the United States.

WHERE ARE ENVIRONMENTAL PERMITS ISSUED?

The U.S. EPA
The U.S. Army COE
State Environmental Resource Agencies

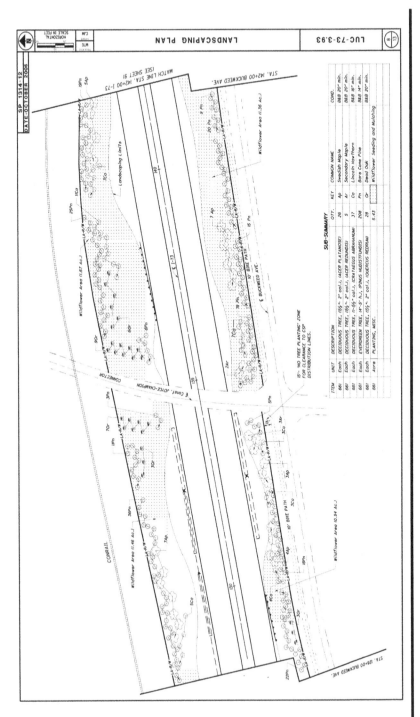

Figure 6.4 Illustration of landscape design plan. (Ohio Department of Transportation, http://www.dot.state.oh.us/Divisions/ Engineering/CaddMapping/sample_plans/Pages/sample_plans.aspx.)

Design plans provide vital information to construction personnel in the field for building a road. Each design plan also contains estimates on quantities of various construction materials needed.

6.1.10 *Standard Pay Item*

Among all materials used in highway construction, the majority are standard construction materials. Standard construction materials are predefined wares with known characteristics, dimensions, applications, and well-understood strengths and weaknesses. Standard construction material is also known as standard pay item. A state highway agency may have thousands of standard pay items. Under normal circumstances, design and construction engineers are required to utilize standard pay items.

STANDARD PAY ITEMS

Clearly defined by state DOTs with specificities such as state, shape, size, and material used.

Advantages of using standard pay items include:

- Standard pay items are well defined and well known.
- Quality is easier to monitor and control.
- Standard pay items are typically cheaper to build and manufacture.
- Standard pay items reduce construction complexity.
- Standard pay items reduce costs for repair, rehab, and replacement during maintenance.
- Standard pay items are easier for State DOTs to manage contracts.
- Standards pay items are easier for contractors to prepare bids.

Table 6.1 illustrates quantity information for standard pay items as part of a roadway plan.

Table 6.1　Standard Pay Item Estimates (Quantities)

Pay Item No.	Description	Unit	Quantity
1246	Wear Course Mix 2C	Ton	65,261
2312	12 × 12 Precast Concrete Box Culvert	Linear Foot (LF)	102
2876	12 × 12 Precast Concrete Box Culvert End Section	Each	6
3127	Granular Backfill	Cubic Yard (CY)	1,864

(Continued)

Table 6.1 (*Continued*)   Standard Pay Item Estimates (Quantities)

Pay Item No.	Description	Unit	Quantity
3366	Common Excavation	CY	2,698
3890	Muck Excavation	CY	1,234
3566	Granular and Select Granular Borrow	CY	1,420
2342	Geogrid	Square Yard (SY)	120
2661	Aggregate Base and Shouldering	Ton	82,899
1766	SP 12.5 Wearing Course Mixture	Ton	49,900
1886	Thermoplastic Pavement Markings	Linear Foot (LF)	36,172

6.1.11 Commitments

From planning to preliminary engineering (PE) and the National Enviornmetnal Polciy Act (NEPA) anlysis to final engineering, a highway agency may have made a host of commitments to the public, property owners, neighborhoods, environmental regulatory agencies, and others. Commitments are promises, and promises need to be kept. Final design engineers are required to ensure all commitments are incorporated into the appropriate design plans. For example, a commitment may be a noise wall for a neighborhood, construction season avoidance, day and night construction noise and light control beyond standard practices, or a beach sand-colored overpass retaining wall and structure.

> **WHY COMMITMENT IMPLEMENTATION IS IMPORTANT?**
> • Meet legal requirements
> • Maintain credibility

Design engineers need to be fully aware of the fact that if a commitment is not incorporated into a project's design plan, it will not be implemented during construction.

6.2 Design Standard and Specification

6.2.1 General Design Specification

Final engineering design follows the same engineering principles and specifications as utilized in the Preliminary Engineering design. State highway agencies develop local state-specific highway and bridge design standards in compliance with AASHTO's "A Policy on Geometric Design for Highways and Streets" specification.

State highway agency specifications are more tailored to local characteristics such as climate, terrain, and driving population.

Figure 6.5 shows example roadway design manuals from the Pennsylvania DOT and the Commonwealth of Virginia DOT. State DOTs offer extensive training to their design engineers on design manual usages and applications.

6.2.2 Bridge Design Specification

There are many ways to categorize bridges: material used, functions served, size, and design methods. From a superstructure design standpoint, bridges can be divided into five major categories, arch bridge, girder bridge, truss bridge, suspension bridge, and cable-stayed bridge (Figure 6.6a).

An arch bridge is a bridge shaped like a curved arch with abutments at each end. Bridge load, which includes its own weight and all other (e.g., vehicles, pedestrians, snow) weights, is transferred to its horizontal thrust restrained by the abutments at either side. There are many types of arch bridges. Arch bridges belong to one of the oldest bridge types practiced as its design can utilize simple material such as stone and brick.

A girder bridge refers to a bridge where its slab is supported by girders. Girders can be constructed with rolled steel girders, plate girders, structural steel box girders, prestressed concrete box girders, or composite reinforced concrete and steel girders. A cross section of the box is typically rectangular or trapezoidal. The load from the slab is transferred to the girder and then to the piers and abutments (Figure 6.6b).

A truss bridge is a bridge where its superstructure is constructed with connected elements in the form of triangular units. Connected elements are typically steel. Loads are transferred to piers and abutments via the interconnected triangular units. Stress responding to certain loads on these connected elements may be from tension, compression, or both (Figure 6.6c).

A suspension bridge is a bridge where its deck is suspended by cables and suspenders. The cables are suspended between towers (cable moves freely across a tower) and anchored at each end of the bridge. Vertical suspenders connect the deck below and the cable above. All loads are transformed into tension in these cables. Given the strength of modern materials, particularly the extraordinary tensile strength of steel, the span width of a suspension bridge can be extremely long (Figure 6.6d).

A cable-stayed bridge is a bridge where its deck is suspended by cables anchored to free-standing towers. The appearance of a cable-stayed bridge may be like that of a suspension bridge, but the load-supporting mechanism is completely different. With a cable-stayed bridge, suspension cables do not move freely across the supporting tower. Instead, these cables are anchored to the tower.

For bridge design, loading (both dead and live loads), seismic consideration, and security are key concerns for designers. For bridges over any water, hydraulic data and information about water flow, scouring, navigation, and geotechnical conditions are also required. During the design phase, structural engineers work closely with hydraulic engineers and geotechnical engineers in designing both the sub- and superstructures (Figure 6.6e).

Figure 6.5 Sample roadway design manuals for the states of Pennsylvania and Virginia. (Courtesy of VA at www.virginiadot.gov and Pennsylvania DOT at www.penndot.gov.)

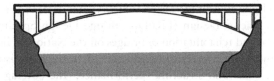

Figure 6.6a   An arch type of bridge illustration.

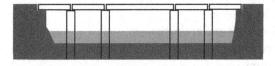

Figure 6.6b   A girder type of bridge illustration.

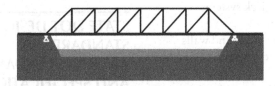

Figure 6.6c   A truss type of bridge illustration.

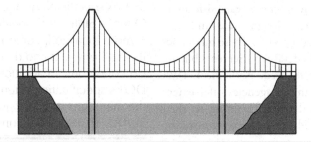

Figure 6.6d   A suspension type of bridge illustration.

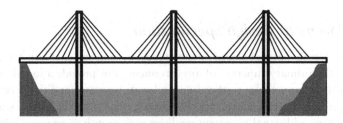

Figure 6.6e   A cable-stayed type of bridge illustration.

The Federal Highway Admnistration (FHWA) mandates the use of AASHTO's "Load Resistance Factor Design (LRFD) - Bridge Design Specifications" for all design, evaluation, and rehabilitation of bridges on the National Highway Systems. The LRFD method relies on information gained through statistical correlation analysis between loads and structural performance. The coverage of the LRFD Specification is broad as illustrated by its 15 chapters listed below:

1. Introduction
2. General design and location features
3. Loads and load factors
4. Structural analysis and evaluation
5. Concrete structures
6. Steel structures
7. Aluminum structures
8. Wood structures
9. Decks and deck systems
10. Foundations
11. Abutments, piers, and walls
12. Buried structures and tunnel liners
13. Railings
14. Joints and bearings
15. Design of sound barriers

STATE DOT DESIGN STANDARDS VS. AASHTO STANDARD AND SPECIFICATION

State DOT design specifications are in compliance with AASHTO/FHWA specifications and criteria. Often, a State DOT specification is more localized, conforming to its local physical and demographical conditions. In addition, a state DOT's specification often exceeds the minimum standard prescribed in AASHTO specifications.

State highway agencies establish more specific structure design guides and specifications (Figure 6.6f). These guidelines conform to Federal requirements but are more locally focused. Both the FHWA and state highway agencies offer extensive structural design training to their employees and private sector consulting personnel.

Figure 6.6f shows the bridge design specifications issued by the State of Georgia DOT and the Washington State DOT.

6.2.3 Pavement Design Specification

Pavement refers to material placed on top of local natural soil enabling smooth vehicle travel. The primary function of any pavement is to provide a long-lasting and smooth surface for vehicles to travel on under all weather conditions. A roadway surface must also provide adequate skid resistance enabling vehicles to stop safely. Over the years, additional pavement qualities such as quiet pavement minimizing tire noise, open drainage pavement facilitating stormwater runoff, and even solar electrical energy producing pavement have been researched and practiced.

Source: Courtesy of Washington DOT at at www.wsdot.wa.gov and Georgia DOT at www.dot.ga.gov

Figure 6.6f Sample bridge design manuals from the states of Washington and Georgia. (Courtesy of Washington DOT at www.wsdot.wa.gov and Georgia DOT at www.dot.ga.gov.)

To achieve the goal of long-lasting and smooth travel under all weather conditions, it is critical that pavement can efficiently distribute vehicle loads to the soil under it. Stress at the wheel pavement contact area should be adequately addressed. After the local stress propagates further down different pavement layers, the stress should not exceed any layer's bearing capacity.

6.2.3.1 Asphalt Concrete Pavement

The deployment of asphalt produced by the petroleum industry with aggregate for roadway pavement is called asphalt concrete pavement or flexible pavement. Flexible pavement typically consists of several layers – surface layer, base layer, subbase layer, subgrade layer (compacted natural soil), and the original soil layer (Figure 6.7a).

The surface layer may be designed and constructed with two sublayers – the wearing course and binder course (also known as intermediate). The base layer is typically constructed with a durable, inert, aggregate that is capable of resisting water and frost/freeze damage. The subbase is the layer between the base and the subgrade. The material used for the subbase is typically inferior to those used in the base layer and can range from less uniform gravel to enforced soil. A subbase's primary function is to provide further structural support to the base. A subbase layer is not always needed or used. A subgrade is the layer directly on top of the natural soil at the site and is typically just a compacted layer of the original or replaced soil.

Binding material is used to promote adhesion between layers. A light application of asphalt, called a tack coat, is applied to the binder course before placing the wearing course. Tack coat application is done by spraying a thin layer of liquid asphalt uniformly on the binding course through an array of nozzles installed on the rear of a tanker truck.

Before laying the binder course or surface course on top of a base layer, another adhesive application called prime coat can be applied to the base layer where it penetrates the base layer through gravity and forms a water-resistant layer. The need for a prime coat depends on local drainage and characteristics of the base layer material.

6.2.3.2 Rigid Pavement

The rigid pavement is also known as Portland cement concrete (PCC) pavement. Its supporting mechanism for vehicle loads is entirely different from that of flexible pavement. While flexible pavement transmits its loads via the interlocking of

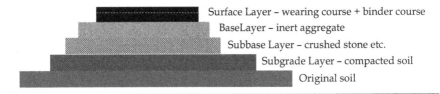

Surface Layer – wearing course + binder course
BaseLayer – inert aggregate
Subbase Layer – crushed stone etc.
Subgrade Layer – compacted soil
Original soil

Figure 6.7a   Flexible pavement layer illustration.

individual aggregates at a highly localized area, rigid pavement transmits its loads over a wide area of subgrade because of its rigidity and high modulus of elasticity. While flexible pavement design is significantly impacted by subgrade strength, rigid pavement design is determined by concrete's flexural strength.

Given PCC's rigidity and high modulus of elasticity to distribute loads over a relatively large area, the base course preparation for rigid pavement is typically less demanding as compared with that for flexible pavement (Figure 6.7b).

The FHWA promotes pavement life cycle analysis in design evaluation. The FHWA also promotes the adoption of the mechanistic-empirical pavement design approach. Historically, pavement design has been empirically driven. The empirical design method relies on historical and experimental data with limited to no quantitative understanding of the mechanics of materials and forces involved.

AASHTO's "Guide for Design of Pavement Structures (1993)" is the primary empirically driven approach used to design new and rehabilitated highway pavement. The output of the method is layer thickness, with computational steps outlined below.

- Specify design period – design life for the pavement per agency policy (e.g., 20 years).
- Estimate design traffic – cumulative traffic during the entire design life expressed in the unit of Equivalent Single Axle Loading (ESAL).

    One ESAL represents the damage done to the pavement by an 18,000-lb single-axle dual-tire once. Total ESAL is the cumulative equivalent damage to the pavement from all vehicles over the entire design period.
- Select a design reliability factor – per agency policy (e.g., 90%).
- Select allowable serviceability loss – maximum loss of roadway surface condition due to traffic before resurfacing is needed – per agency policy.
- Compute seasonally averaged subgrade resilient modulus – elastic property of soil – based on soil core data.
- Determine structural layer coefficients – bound layers vs. unbound layers.
- Determine drainage coefficients – unbound layers only.
- Compute structure number (SN).
- Determine the thickness of various layers in sequential order.

The mechanistic-empirical pavement design method is based on a significant amount of research over several decades, including the Long-Term Pavement Performance (LTPP) program managed by the FHWA. This new method provides flexibility in

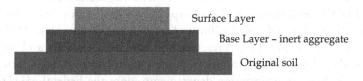

Surface Layer

Base Layer – inert aggregate

Original soil

Figure 6.7b   Rigid pavement layer illustration.

WHY DOES FLEXIBLE PAVEMENT REQUIRE MORE INTENSE BASE LAYER PREPARATION THAN RIGID PAVEMENT

Flexible pavement relies on local interlockings of aggregates to transmit loadings into sublayers. Rigid pavement relies on concrete's high modulus of elasticity to distribute loads over a relatively large area where stress on the sublayer on per square inch basis is much smaller for sublayer as compared to the flexible pavement case.

---

analyzing highly customized pavement layers to meet performance specifications. The mechanistic-empirical pavement design method can be used to analyze a wide range of scenarios, such as:

■ Conventional flexible pavements
■ Deep strength flexible pavements
■ Full-depth hot-mix asphalt
■ Hot-mix asphalt overlay
■ Jointed plain concrete pavement
■ Continuously reinforced concrete pavement
■ Jointed plain concrete pavement overlay
■ Continuously reinforced concrete pavement overlay

The mechanistic-empirical pavement design software developed by AASHTO analyzes pavement layer scenarios through trial runs. Design engineers supply initial layer thickness data for a set of specified design criteria. The software analyzes various stresses associated with the initial layer data supplied by designers. The output is the roughness of the pavement for the trial layer scenario. From there, designers can adjust the layer data (scenario) until an acceptable outcome is obtained.

Regardless of methods or approaches, pavement engineers' knowledge and local experience are always needed in order to assess the reasonableness of any computed values.

Pavement design is carried out by pavement engineers. Traffic, soil, and other geotechnical data are typically provided to pavement engineers.

6.2.4 Pavement Marking

Under the authority of Title 23 Code of Federal Regulation Part 655, Subpart F, the FHWA working with the transportation community establishes and publishes the "Manual on Uniform Traffic Control Devices" (MUTCD) (Figure 6.8a). The MUTCD prescribes mandatory standards and specifications for all public agencies to follow concerning roadway traffic control signs and devices. State transportation departments adopt the Federal MUTCD via State codes.

Pavement marking is the application of colored adhesive and other material to pavement surfaces aiding drivers in navigating appropriate travel paths. The most common pavement markings are solid or dashed, white or yellow lines. White lines indicate the same directional travel for lanes separated by them; yellow lines signify opposite directional travel on the other side of the line.

Pavement markings can also be achieved with reflective mirror devices embedded in the pavement following the path of adhesive tapes or paints with a regular interval. Typically, these reflective devices have two exposed angled mirrors. When a vehicle's headlight shines on these mirrors in the direction of travel, these mirrors have a white glow signifying lane boundary. When an opposing headlight beam shines on them, these mirrors have a red or yellow glow, warning drivers they are traveling in the wrong direction.

A pavement marking plan shows where delineation and markings should be placed. Construction personnel follows the marking plan to stripe or install devices accordingly.

Figures 6.8b and 6.8c illustrate how pavement markings are used to control lane usage.

Figures 6.8b and 6.8c show how critical pavement markings are. Without them, drivers would not be able to navigate a roadway efficiently and safely.

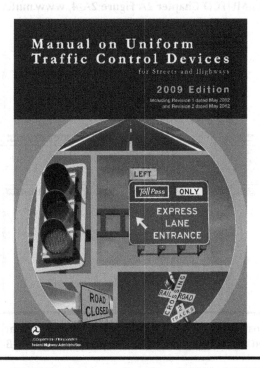

Figure 6.8a The MUTCD. (Courtesy: Based on FHWA's MUTCD https://mutcd. fhwa.dot.gov/.)

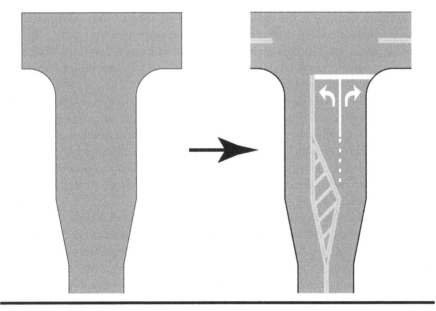

Figure 6.8b **Illustration of pre- and post-pavement marking effects. (Courtesy: Based on FHWA's MUTCD Chapter 2A Figure 2A-4, www.mutcd.fhwa.dot.gov.)**

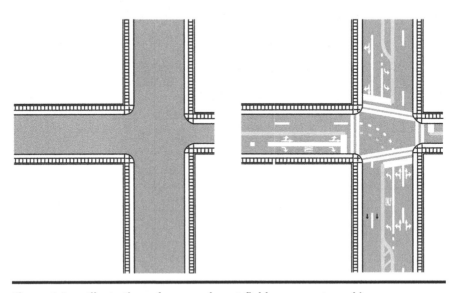

Figure 6.8c Illustration of pre- and post-field pavement marking appearance. (Courtesy: Based on FHWA's MUTCD Chapter 3B Figure 3B-27, www.mutcd. fhwa.dot.gov.)

6.2.5 Roadway Signs

Roadway signs are static or dynamic instructional displays directing and informing travelers for in-situ decision-making. Signs include displays placed on a pavement, overhead structures, and roadside posts within the right of way. All highway agencies are required to follow the standards contained in the MUTCD.

Sign placement shall consider:

- Types of sign – regulatory, warning, guidance, information, and others
- Material of signs
- Size of signs
- Lettering – font, size, and color
- Location of signs including advanced locations

The placement of signs is to inform and enable effective decision-making by drivers with an appropriate amount of information. Too much information overwhelms drivers, while too little would not be informative enough. Additionally, sign uniformity is critical for meeting driver expectations.

In scenarios where the standard specification may not work as needed, design engineers shall carry out a site-specific study by following the principles prescribed in the MUTCD to develop a localized sign plan.

Figure 6.9a illustrates signing for a two-lane intermediate or minor interchange exit according to the MUTCD. Figure 6.9b illustrates the signage of a typical intersection approach, according to the MUTCD.

Figures 6.9a and 6.9b show how roadway signs are placed to inform drivers for their decision-making. These signs enable driers to make appropriate decisions when faced with multiple options.

6.3 Design Exception

Roadway design speed, horizontal curvature, vertical clearance, lane width, shoulder width, superelevation rate, grade, cross slope, design loading, structural capacity, stopping sight distance, border width, clear zone, and lateral offset to obstructions are all design elements. Design engineers are required to meet standard specifications for these elements.

Occasionally, design engineers may not be able to meet a design element's standard design criterion. In such cases, a design engineer may seek a variation from the established standard. The departure from the standard design specification is called a design exception. Seeking a design exception should be taken seriously. Appropriate justification, analysis, and approval before implementing any design exception are needed. Design engineers should use design exceptions as a last resort. When a design exception is considered, it should proceed with great caution.

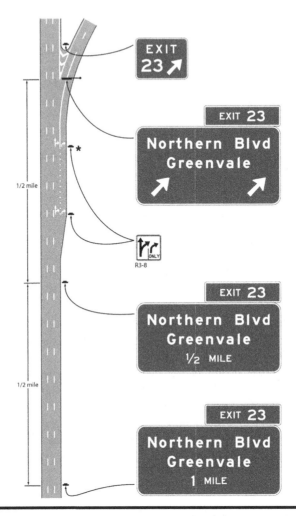

Figure 6.9a   Signing a two-lane intermediate or minor interchange exit. (Courtesy the FHWA MUTCD Chapter 2E Figure 2E-12, www.mutcd.fhwa.dot.gov.)

A design engineer's design exception analysis typically covers a minimum of the following areas:

■ existing roadway characteristics
■ alternative solutions considered
■ safety impact
■ operational impact
■ environmental impact
■ mitigation needs
■ cost impact

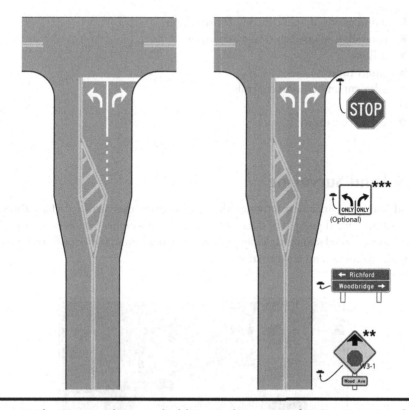

Figure 6.9b   How to sign a typical intersection approach. (Courtesy: Based on FHWA's MUTCD Chapter 2A Figure 2A-4, www.mutcd.fhwa.dot.gov.)

6.4 Constructability Review

A constructability review is a project management process to ensure that an engineering design can be built with standard methods, normal equipment, and standard techniques. The constructability review ensures that a project will be biddable and offers the best value for the project owner.

A constructability review starts at the preliminary engineering stage and becomes more focused as the project moves into its final design. During the constructability review, it is critical that personnel (not design engineers) with extensive field construction experience review the concept and plan, thus ensuring the project is buildable, biddable, maintainable, and cost-effective.

Typical issues considered under constructability review include:

■ clearing/grubbing
■ removal/demolition structures
■ nature and environment protection methods
■ stormwater management problems during phased construction

- maintenance of traffic during construction
- materials and geotechnical exploration
- utilities
- detour routes
- sediment and erosion controls
- drainage
- reconstructability

6.5 Land Surveying

Land surveying includes activities that characterize the earth's surface through three-dimensional geospatial data, covering point positions, distances, and angles. Land survey, as related to highway projects, includes boundary identification, topographic mapping, and construction staking.

An integrated modern electronic transit theodolite capable of electronic distance, vertical angle, and horizontal angle measurements, coordinate determinations, and onboard computer driven data collection and triangulation calculations. Such a device is also called a Total Station as it can independently make a complete set of map.

6.5.1 Control Surveys

Control surveying is to establish precise vertical and horizontal reference monuments. Horizontal control surveying establishes geodetic latitude and longitude information covering a large area. Within that area, a rectangular plane coordinate in the form of a state plane coordinate system is commonly used to record

coordinates of all points. Vertical control surveys are activities that determine elevations based on reference monuments.

The most common method for a horizontal control survey is called triangulation, while the most common practice for the vertical control survey is differential
leveling. The National Geodetic Survey (NGS) established a nationwide horizontal
control framework with more than 270,000 stations in 1983. This horizontal control
framework is called NAD83. The NGS also established a network of monuments
with their elevation information identified against the mean sea level. The 1988
North American Vertical Datum known as NAVAD88 provides the latest extensive
vertical reference framework.

6.5.2 Right of Way Surveying

Right of way surveying demarks land boundaries based on land ownership.
Boundary surveying is the oldest and most demanded type of survey. For highway
right of way surveying, survey work provides information to establish the project boundary, flag and stake right of way boundary in the field, resolve boundary
disputes, and create legal descriptions. After all, transportation agencies need to
ensure that their project activities are carried out on land owned by agencies or
permitted by owners through easements.

6.5.3 Engineering Design Surveying

Engineering design surveying encompasses survey activities collecting topographic
data for highway projects. The topographical data enables a complete mapping of
land surface with a three-dimensional coordinate system (latitude, longitude, and
elevation). Both conventional (on the ground) and photogrammetric (in the air)
methods are used in an engineering design survey.

The engineering design survey generates a triangulated irregular network (TIN)
data consisting of various vertices (points) connected by their edges to form a triangular tessellation (face). Figure 6.10 provides an example of the TIN file data

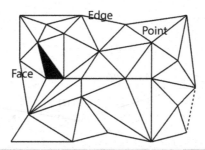

Figure 6.10   A sample plot of triangulated irregular networks (TIN).

plot. Design engineers rely on the TIN file to establish both horizontal and vertical alignments, minimizing the right of way needed, soil transport (cut and fill), and impacts to both human and natural environment.

6.5.4 Construction Surveying

Field construction relies on stakes showing lines and grades. The construction surveying transfers design specifications (e.g., alignment and boundary) from a design plan to the ground by establishing reference lines and points.

A staking process typically starts at the point where the first tangent segment begins, proceeds with 100-feet (one full station) interval until the alignment changes its direction at the first point of intersection (PI). By following the deflection angle, the second tangent is established. Stake the second tangent until the alignment changes its direction again. Continue this process until it reaches the alignment's terminus, as illustrated in Figure 6.11. Once all the tangents are staked out, horizontal curves are fitted for all PIs.

A wide variety of survey stakes are placed to guide construction activities. These construction survey stakes include but are not limited to:

- clearing stakes
- slope stakes
- fence stakes
- rough grade stakes
- final grade stakes
- drainage stakes
- curb stakes
- structure stakes
- cut and fill stakes

Points (stakes) used for referencing must stay outside construction limits to avoid destruction and alteration during construction. Referencing stakes are critical, given that centerline of construction stakes will most likely be destroyed during construction.

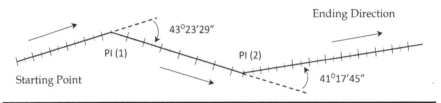

Figure 6.11   Staking out a roadway alignment by following tangents.

6.5.5 Staking a Horizontal Curve

A procedure called defection and cord method is one of the common approaches used to stake out a horizontal curve. In this example, the work is to stake out a 2° curve illustrated in Figure 6.12a

Steps:

1. Set the theodolite over the PC point (A). As illustrated in Figure 6.12a, the PC is at Station 270+35.25.
2. Locate the next full station following the PC station on the curve. Compute the arc length between the two stations.

    In this example, the next full station is Station 271+00, which results in a curve length (arc length) between the two stations to be: (271 × 100 − 270 × 100 − 35.25) = 64.75 feet.
3. Compute the central angle (c) corresponding to the arc length by using the formula

$$c = \frac{360L}{2\pi R}$$

    where c is the central angle in degree, L is the arc length in feet, and R is the curve radius in feet.

    For this example, the curve is 2°. The 2° curve results in a 2,866-feet radius. The central angle is (360 × 64.75)/(2 × 3.14 × 2,866) = 1.295° = 1° 17′42″. The chord length (CL) between the PC and Station 270+35.25 is CL = 2 × 2,866 × sin (1.295°/2) = 64.71 feet.
4. Compute both the central angle and chord length corresponding to a full station (100-feet arc length).

    For the above example, for a full station, the central angle is (360 × 100.00)/(2 × 3.14 × 2,866) = 2.002° = 2°0′36″. The chord length is CL = 2 × 2,866 × sin(2.002/2) = 99.98 feet.
5. Stake the first station on the curve ahead of the PC station.

    In the example, the goal is to stake Station 271+00. A backsight is taken on the PI (point of intersection) with the instrument's horizontal circle set to zero.

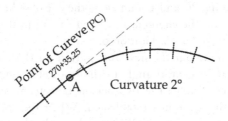

Figure 6.12a Centerline of construction.

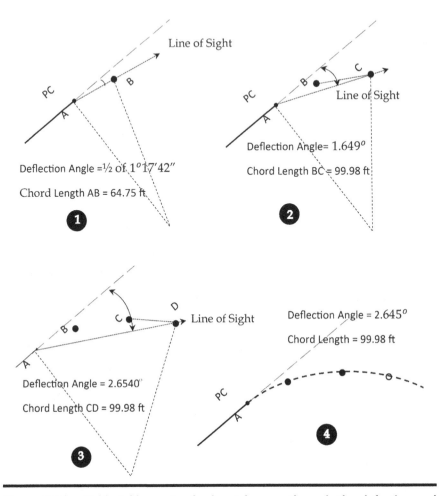

Figure 6.12b Field staking out a horizontal curve through the defection and chord method.

Turn the instrument clockwise to a defection angle of half of the central angle, which is ½ of 1°17′42″. A two-person tape team measures the cord length of 64.75 between the PC and a point along the line of sight from the instrument (Figure 6.12b_1). The endpoint of the 64.75 feet in the line of sight of the instrument is staked out as location B.

6. Continue to stake out the next full station on the curve by adding half of the full station central angle to the last defection angle and turn the theodolite accordingly. The two-person tape team measures a full cord length from the stake just established. When the endpoint of the tap is aligned with the instrument's line of sight, stake the station on the ground accordingly.

In this case, the new total deflection angle is 1.295°/2 + 2.002°/2 = 1.649° (Figure 6.12b_2). The two-person tape length is 99.98 feet from B. Mark point C where the 99.98-feet endpoint appears in the line of sight of the theodolite.

7. Continue the process until all stations are staked out based on deflection angle and chord length. In this example, the continued marking of point D is shown in Figure 6.12b_3.

8. Construction personnel will follow stakes A, B, C, D, etc., to perform field operations (Figure 6.12b_4).

After a centerline of construction or other reference line is established, stakes marking the boundary are placed. The placement of right of way stakes is typically done by measuring the perpendicular distance from the reference line outward.

6.5.6 Labeling Stakes

Survey stakes are pieces of wood that come in a variety of dimensions such as the common 18 inches × 2 inches × 1 inch (H × W × L) specification. Labeled stakes convene critical information to field construction personnel. Although there are some standard practices for labeling and marking stakes, each state transportation department may have its unique specifications. It is critical to recognize a project's jurisdiction and interpret the respective construction stakes accordingly.

Figure 6.13a shows a common clear and grubbing stake in the field. On one side of the stake, it is marked with the words of "Clearing Limit," indicating that clearing shall not exceed this point. The Station 175+60 marked on the other side

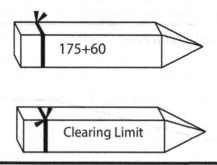

Figure 6.13a   Clearing stake illustration.

Clearing and Grubbing Limit Stake

The front of the stake has the station written on it corresponding to the design plan. The back of the stake has the clearing limit written on it.

Construction crew shall not clear grounds beyond this point.

of the stake corresponds to the construction plan, providing the context to the land clearing and grubbing activity.

Figure 6.13b illustrates another type of stake known as the slope stake. Equipment operators and field personnel can rely on this stake to make the cut and fill as indicated. The label on this stake leads to the result illustrated in Figure 6.13c, as explained below from Steps 1 to 5.

Steps:

1. C-BS 3′@2:1

   Cut (C) a back-slope (BS) 3-feet deep with a 2:1 slope – resulting in 6-feet wide (horizontally) land strip forward.

2. 3′ D.B.

   Grade (level) a 3-feet ditch-bottom (D.B) – resulting in another 3-feet strip of land forward.

3. F.S. 4′@4:1

   Grade the fore-slope (F.S.) 4-feet high with a 4:1 slope – resulting in 16 flat feet of land forward.

4. 3′SHD@3%

   Grade 3 feet at a 3% slope for the shoulder (SHD) – resulting in approximately 3-feet strip of land forward.

5. 13′CL@2%

   Grade 13 feet at a 2% slope toward the centerline (CL) – resulting in a 13 flat feet strip of land forward.

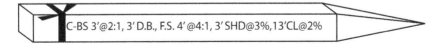

Figure 6.13b   Slope stake illustration.

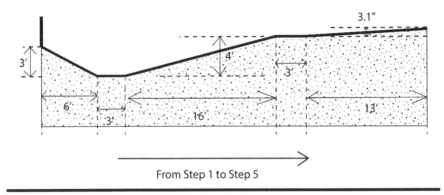

From Step 1 to Step 5

Figure 6.13c   Cross section produced as a result of executing the slope stake.

6.6 Next Step

When a set of the final engineering design plan is complete, the project can move into the right of way acquisition phase if any additional right of way is needed. For a project where no additional right of way is needed, the project can move into the construction phase directly (e.g., bridge projects over water often do not need any additional right of way).

6.6 Summary

The final engineering continues the preliminary engineering's design on both the vertical and horizontal alignments. The final engineering completes all the design details needed and delivers the plans (P), specifications(S), and estimates (E) known as the PSE package.

The final design follows the exact design specifications and design standards established by state DOTs as in the preliminary engineering design. These state-specific design standards and criteria are often more demanding and more stringent than what are specified in AASHTO's "A Policy on Geometric Design for Highways and Streets" and other AASHTO specifications.

Occasionally, standard design specifications and criteria can't be met. In such a case, a design exception is sought. However, before a design exception can be implemented, a host of factors such as safety, cost, and environmental impact must be fully analyzed and documented. A design exception request package is prepared and submitted for approval following agency established procedures.

Land surveying provides information on property boundaries and topography. The right of way surveying offers specific property boundary information. The design surveying contains both boundary data and surface elevation data. And the construction surveying transfers all dimensional data from a set of design plans to the field where construction personnel can reference.

One critical aspect of the final engineering design is to ensure all commitments made during planning, EIE, PE, and the final design are included in the design plan. If a commitment is not part of the PSE package, it will not be implemented during construction.

6.8 Discussion

1. By State statutes, virtually all civil engineering design plans need to be signed and sealed by a professional engineer registered in the State. Briefly discuss how such laws and regulations protect the public interest and promote the engineering profession.

6.9 Exercises

1. Final engineering design is the continuation of the preliminary engineering design that has been approved by the lead Federal agency during the environmental impact evaluation and PE stage. True or False

2. The final engineering design delivers three key products – the design plan, the specification, and the quantity (estimates) of material. The three components make up the so-called PSE package. True or False

3. Identify potential components of a set of roadway construction plans from the list below.
   a. Roadway design plan
   b. Structure design plan
   c. Portland cement manufacturing plan
   d. Pavement marking plan
   e. Landscape plan
   f. Street lighting plan
   g. Asphalt refinery production plan

4. Identify potential permits associated with a roadway construction project from these listed below.
   a. Vehicle emission permit
   b. NPDES – construction permit
   c. Wetland permits (Section 404 permits)
   d. Highway noise abatement wall permit
   e. Bridge permit

5. What is a standard pay item for highway construction projects?

6. What are the advantages of using standard pay items for highway construction projects?

7. Where are commitments typically made and recorded prior to the final engineering design?

8. On the national level, there are design specifications called "A Policy on Geometric Design for Highways and Streets" and the "Load Resistance Factor Design - Bridge Design Specification." The national specifications represent the minimum requirements to be carried out by all agencies involved in highway and bridge work. True or False.

9. State DOTs have their local roadway design standards and specifications to guide local decision-makings. Often a state DOT standard and specification are more stringent than the national specifications. True or False.

10. State DOTs, the FHWA, and private institutions offer a wide range of roadway and bridge design training classes to both their own employees and professionals interested in the subjects. True or False.

11. List bridge types based on superstructure design.

12. There are two main types of pavement – asphalt concrete (flexible pavement) and Portland cement concrete (rigid pavement). True or False.

13. Flexible pavement design requires more preparation for its base and subbase as compared to rigid pavement. True or False.

14. The surface layer of asphalt pavement may consist of two layers – a wearing course and a binder course. True or False.

15. To ensure pavement wearing course adherence to the underlying binder course, a thin liquefied layer of asphalt known as tack coat is often applied to the binder course before placing the wearing course. True or False

16. What does ESAL stand for?

17. The ESAL approach in pavement design is often referred to as empirical, where minimum stress and strain analyses are used in the design computation. True or False.

18. Both the FHWA and AASHTO promote the adoption of the mechanistic-empirical pavement design (MEPD) method for pavement design. True or False.

19. Pavement marking and roadway sign design need to follow specifications prescribed in the Manual on Uniform Traffic Control Devices (MUTCD). True or False.

20. The MUTCD's specification for roadway signs is highly prescriptive. Material, color, lettering size, font, etc., are all detailed. True or False.

21. The design exception process offers a path to deviate from standard specifications. True or False

22. The constructability review during design ensures that the final design is biddable and constructible. True or False.

23. Land surveying offers data and information on property boundaries, location, distance, angle, and surface characteristics. True or False.

24. Land surveying is a regulated profession meaning that personnel must be licensed by appropriate governmental regulatory agencies to perform such activities. True or False.

25. NAD83 is the national horizontal survey reference monument network. True or False.

26. NAVAD88 is the North American Vertical Datum reference network. True or False.

27. Design engineers need surface elevation data to analyze the needs of cuts and fill in order to design a smooth vertical alignment. True or False.

28. Design engineers rely on surveyors to provide surface elevation data. Such data is typically transmitted to design engineers in a TIN file. True or False.

29. Construction surveying establishes needed reference marks and lines such as grades, boundaries, and limits information for field personnel to reference and abide by. True or False.

30. During a construction surveying, staking out a horizontal alignment typically starts with laying out tangents first and then adding horizontal curves to the alignment. True or False.

Bibliography

A Policy on Geometric Design for Highways and Streets, 2018, AASHTO.

Bridge and Structures Design Manual, Revision 2.3., Georgia Department of Transportation, 7/11/2017, http://www.dot.ga.gov/PS/DesignSoftware/Bridge.

Bridge Design Manual (LRFD), M 23-50.17, June 2017, Washington State Department of Transportation, https://www.wsdot.wa.gov/Publications/Manuals/M23-50.htm, https://www.wsdot.wa.gov.

Design Manual Part 2- Highway Design, Publication 13M, Pennsylvania Department of Transportation, March 2015, https://www.dot.state.pa.us/public/Bureaus/design/PUB13M/Chapters/TOC-2.pdf, https://www.dot.state.pa.us.

Guidance on NHS Design Standards and Design Exceptions, FHWA, https://www.fhwa.dot.gov/design/standards/qa.cfm.

Guide for Design of Pavement Structures, 4th Edition with 1998 Supplement, AASHTO.

LRFD Bridge Design Specifications, 8th Edition, AASHTO.

Manual on Uniform Traffic Control Devices (MUTCD), 2009, Federal Highway Administration, https://mutcd.fhwa.dot.gov/.

Mechanistic-Empirical Pavement Design Guide: A Manual of Practice, 2nd Edition, AASHTO.

Road Design Manual, Virginia Department of Transportation, Issued January 2005, Revision, July 2017, http://www.virginiadot.org/BUSINESS/locdes/rdmanual-index.asp, http://www.virginiadot.org.

Sample Plan Sheets, Ohio Department of Transportation, http://www.dot.state.oh.us/Divisions/Engineering/CaddMapping/sample_plans/Pages/sample_plans.aspx, http://www.dot.state.oh.us.

Sample Roadway Cross Section Sheet, Maryland Department of Transportation, MDDOT, Maryland Open Bid, https://emaryland.buyspeed.com/bso/external/bidDetail.sdo?bidId=MDJ0231027399&parentUrl=a.

Chapter 7

Right of Way

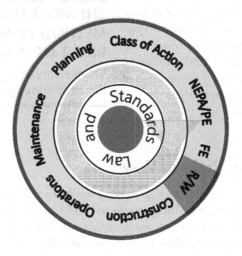

The concept of ownership forms the foundation of both ancient and modern societies because it guarantees certain rights. Guarding a person's property and possessions is typically viewed as one of the most sacred laws of justice.

In the United States, property (land, buildings, and other improvements) ownership and ownership transfer information are legally recorded by county governments per state statutes. The Federal government does not possess modern-day real property transactions and ownership records.

A final roadway engineering design plan identifies all additional lands needed for a project. Before any construction can begin, it is required that all the needed lands are secured or will be secured with an identified and executable plan. Starting a field construction without acquiring all needed lands often results in a costly budget overrun and delayed schedule.

7.1 Right of Way Definition

Right of way (ROW) is a legal right to create a way through a piece of land. To have this right, a party can either be the owner of the land (title holder) or has legal permission from the owner of the land. The easiest way to gain legal land ownership is through purchasing. When a piece of land cannot be purchased for its usage right by another party, the other party may still gain a right to use the land for a specified purpose – with or without condition by gaining its landowner's permission. A land owner's permission for others to use his or her land is called an easement. Depending on the specifics of an agreement, an easement can be permanent or temporary.

WHERE PROPERTY RECORDS ARE KEPT

Virtually all modern-day property transaction records are kept and maintained by County governments.

Property ownership is determined by what is recorded in the official County record and not personal recordings.

THE MOST USED METHOD IN ACQUIRING RIGHT OF WAY

The fee-simple purchase method is the most common approach.

7.2 Acquiring Right of Way

The ROW can be acquired through various mechanisms. A fee-simple purchase, which is based on a seller's willingness to sell and a buyer's desire to buy, is most common. Other methods such as owner donations and easement also occur. For highway projects or other public projects, occasionally, a process known as condemnation is used. Condemnation is a legal process where a government agency acquires properties from unwilling and uncooperative owners. The government's eminent domain authority enables a government to acquire private property for public project purposes. During the eminent domain condemnation process, a government unit goes to a court to acquire the needed property.

7.3 Acquisition Procedure

The concept and practice of selling and purchasing form the foundation for our economic system. The critical component for government agencies in acquiring needed ROW is its fairness to both property owners and the government agencies. To ensure such fairness, governments at all levels establish their own procedures to follow. On the Federal level, Congress passed the Uniform Relocation Assistance and Real Property Acquisition Policies Act of 1970 (The Uniform Act), which states that Federal procedures must be followed when federal funds are used for a project.

Federal legislation supersedes any other state or local laws on projects receiving Federal funds. And the definition of receiving Federal funds by a project covers all phases of a project including the ROW acquisition. For example, if a project gets and uses Federal fund for planning, the project's ROW acquisition process must follow the Uniform Act. Also, if a project plans to use Federal funds for the upcoming construction, its ROW acquisition must follow the Uniform Act.

7.3.1 Involving Affected Property Owners

At the earliest stage possible, such as the environmental impact evalution (EIE) and preliminary engineering (PE) phase, responsible transportation agencies should notify adjacent property owners on potential impacts on their properties and seek their inputs.

7.3.2 Official Notification

If a property, or a portion of a property, is required for a project per the final engineering design plan, the property owner shall be notified that the transportation agency is interested in purchasing the property and provided with detailed information on the upcoming property appraisal.

FEDERAL REAL PROPERTY ACQUISITION LAW

The Uniform Relocation Assistance and Real Property Acquisition Policies Act.

7.3.3 Official Appraisal

The official appraisal is a two-stage process. First, agency staff or authorized contractor representatives of a transportation agency coordinate with a property owner to inspect the property and perform an evaluation of its value – known as appraisal. The property owner or the property owner's representative can accompany agency staff or agency representatives to ensure unique features of the property are recognized and accounted for. The end product of the first stage is the determination of a property's fair market value, which is a price that a willing and informed seller and buyer can agree.

Upon the completion of the initial appraisal stage, the agency will have another appraiser perform a control review of the initial value determination. Based on the review of the original appraiser's recommendation on valuation, the agency will determine the final offer, which is called just compensation.

7.3.4 Written Offer

After just compensation is determined, an agency will make a written offer to the property owner. The offer includes the price (purchasing price) the agency is willing to pay for the described property – including all improvements to the property associated with the property, agency procedure, time offered to the property owner to consider, and how to ask additional questions.

7.3.5 Purchasing

Once an agreement is reached between an agency and a property owner with regard to an agency offer, a settlement will be conducted with the property owner signing the agreement. Upon signing the agreement by both parties, the property owner is paid, and the property ownership is transferred to the agency. The transportation agency records the transaction and the new boundary internally. The transaction and new ownership of the property, which is the agency, are also recorded in the local County recording office for official record-keeping.

7.3.6 Mediation or Condemnation

If a property owner does not agree with the agency offer, the agency can choose to use mediation to help to reach an agreement. After the mediation process, the final offer may be different from the initial just compensation offer.

If the agency believes that mediation will not lead to an agreement, the agency may proceed with its eminent domain procedure, where the agency will take the property owner to court and have the court make the value determination. The value determined by the court will be considered the new just compensation. The court-determined just compensation may be higher or lower than the original just compensation determined by the transportation agency.

WHAT JUST COMPENSATION IS

Just Compensation is the offer value determined by the governmental agency by following an established procedure.

Design engineers are often required to testify on why a particular piece of property is required for a project during eminent domain proceedings. Design engineers provide logical and reasonable explanations to the court on why a property is needed.

7.4 Relocation

When a property is acquired for a transportation project, individuals, families, and businesses residing in buildings on the property are displaced. The Federal Uniform Act establishes certain benefits and offers certain protections for people and businesses displaced by Federal projects.

For displaced people and businesses, relocation assistance covers moving expenses and supplemental relocation assistance due to increased costs from departing the original location.

TENANTS VS. OWNERS

With regard to relocation, the Uniform Act does not differentiate tenants from property owners.

Protections offered by the Uniform Act require that the public agency ensures the availability of replacement housing and minimum standards for such housing can be met within a reasonable geographical area.

7.5 Final Deliverable

At the conclusion of the ROW phase, the goal is to have taken possession of all required ROW. Occasionally, due to schedule needs, a feasible plan on how to acquire the remaining ROW is prepared as opposed to completing all actual acquisitions. In these cases, agents from a state transportation department's ROW office need to certify the status of the ROW before a project can move to the next phase if the project is Federally funded.

A state transportation department's ROW office also prepares a set of records and records them with appropriate County for official record-keeping.

7.6 Next Step

Upon acquiring all needed ROW, agents at State Department of Transportation (DOT)'s ROW Office certify that all needed ROW is acquired. The project proceeds to the construction phase.

7.7 Summary

Federal, State, and local governments have their own unique ROW acquisition laws, regulations, and procedures. These regulations and procedures may vary significantly. However, for any project receiving Federal financial assistance, the project's ROW acquisition and relocation procedures must follow the Federal Uniform Act where fair and just compensation must be offered to affected property owners.

Occasionally, property owners may not be willing to sell their properties to governments. If a government entity has determined that negotiation is unlikely to convince a property owner to sell, the government unit can take the property owner to court and force the property owner to sell the property. This legal process is called condemnation, and the power the government exercises is called eminent domain authority.

7.8 Discussion

1. No private citizen or private business can force a person to sell his or her real property, and nor can any private citizen or businesses take a person to court and have the court force the person to sell his or her property. However, all

governments – Federal, State, County, and city, can force a citizen to sell his or her property. The power governments relied on is known as the "eminent domain" power. Private citizens do take government entities to courts to protect their land from being taken by the government. As an engineer, how can you help to protect individual property owner's rights in developing public projects?

2. Outline the relationship between real property acquisition and relocation assistance.

7.9 Exercises

1. Guarding a person's property and possession is typically viewed as the most sacred law of justice in most societies. True or False

2. Transportation agency's right of way agents rely on the following to acquire additional right of way needed for a transportation project. Select all apply.
   a. Preliminary engineering report
   b. Proposed right of way map
   c. Final engineering design plan
   d. Landscape plan
   e. Street lighting plan
   f. Pavement marking plan
   g. Existing agency right of way maps
   h. County government real property records

3. A real property easement from a property owner is to give the land away without any compensation. True or False.

4. A real property easement from a property owner is the owner's permission for someone else to use the property for certain activities with or without compensation. The ownership of the property does not change. True or False.

5. The most popular and the easiest way to acquire additional real property is through the fee-simple method. True or False.

6. A fee-simple real property acquisition is the transaction between a willing seller and a willing buyer where the seller and buyer agree on a transaction price. True or False.

7. All governmental agencies have their unique right of way acquisition (real property acquisition) laws, regulations, and procedures. True or False.

8. For Federally sponsored projects, the right of way acquisition must follow the Federal Uniform Relocation Assistance and Real Property Acquisition Policies Act. True or False.

9. According to the Uniform Act, a transportation agency can simply make an oral offer to a property owner. True or False.

10. A State transportation department needs some land from an unwilling and uncooperative owner to correct a roadway's safety deficiency. What mechanisms does the State Department of Transportation have to get the land it needs?

    a. Build on it when the owner is not around

    b. Offer the owner more money

    c. Threaten the owner with state police and force the owner to sell

    d. Go to court using the eminent domain power

11. A county transportation department plans to use federal funds for a road's construction, county fund for it's right of way acquisition, and local sale tax money for its maintenance. The County does not need to follow the Federal Uniform Act for relocation. True or False.

12. Mr. Smith owns a two-story residential building with four units. Mr. Smith rents all four units out to four separate families. Now his state Department of Transportation notifies him that the State is planning to acquire his entire property to widen the existing roadway in front of his house. Mr. Smith is pleased with the news and is eager to sell his property to the State (Mr. Smith has been having a hard time finding tenants paying rents on time). Mr. Smith told the State that he is ready. Is it true that the State DOT does not need to consider any compensation for his tenants because these tenants do not own anything the State will purchase? Under the Uniform Act, does the State have any legal obligation to assist Mr. Smith's tenants?

13. What is just compensation?

14. Describe what condemnation is as related to real property acquisition by government agencies.

15. Construction typically only occurs when needed right of way is acquired. If a state agency is planning to use Federal fund for its upcoming construction, does the state Agency need to follow the Uniform Act when purchasing needed right of way? Yes or No.

Bibliography

Uniform Relocation Assistance and Real Property Acquisition Policies Act of 1970, Public Law 91-946, https://www.govinfo.gov/content/pkg/STATUTE-84/pdf/STATUTE-84-Pg1894.pdf.

Acquiring Real Property for Federal and Federal-aid Programs and Projects, Revised November 2018, Publication FHWA-HEP-19-010, https://www.fhwa.dot.gov/real_estate/uniform_act/acquisition/real_property.cfm.

Your Rights and Benefits as a Displaced Person under the Federal Relocation Assistance Program, Revised October 2014, Publication No. FHWA-HEP-05-031, https://www.fhwa.dot.gov/real_estate/publications/your_rights/.

General Acquisition and Relocation Information, FTA, https://www.transit.dot.gov/funding/grant-programs/capital-investments/general-acquisition-and-relocation-information.

Chapter 8

Construction

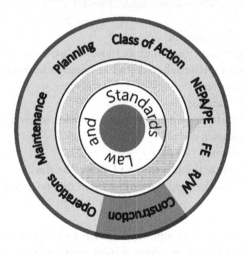

Only after a set of design plans is drawn, needed right of way is acquired, and necessary environmental permits are obtained, is a project ready to be constructed. Successfully implementing a design plan during the construction phase relies not only on the quality of the design plan, but also the experience and knowledge possessed by construction personnel.

Typical steps involved with a roadway construction activity consist of the five major steps listed below.

1. Construction Surveying

   The first step associated with a roadway construction project is construction surveying. Construction surveying places reference points and markings to identify and mark boundaries, referencing and controlling horizontal alignments and vertical grades.

2. Initial Earth Work

The second step is to perform filling and cutting along a horizontal alignment to establish a rough grade for the eventual final vertical alignment.

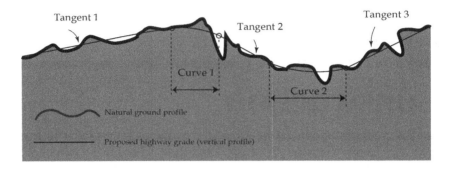

3. Refine Vertical Alignment

During this step, the rough cut and fill work is completed and further smoothed. Appropriate subgrade soil (original or borrowed) is placed per the design specification.

4. Base Layer Preparation

During base layer preparation, gravel or other material is spread to improve load-bearing capacity.

5. Asphalt Placement

The last step is the placement of either the flexible bituminous concrete pavement mix or the rigid Portland cement mix on top of the improved base layer.

In this chapter, five critical components related to construction are covered: personnel, contract administration, construction engineering and inspection (CEI), construction dispute resolution, and construction safety.

8.1 Construction Personnel

To carry out any construction activity efficiently, a wide range of skilled personnel with appropriate experience, licenses, permits, and certificates are needed.

8.1.1 Land Surveyors

Land surveying is a regulated professional activity where practitioners must be licensed by a regulatory board (e.g., Board of Professional Surveys). The regulatory board is sanctioned by its State government through a state statute.

Personnel led and supervised by certified and licensed survey professionals can perform construction surveys. Construction surveying transfers locations and dimensions from a design plan to the ground by establishing reference lines and points such as (a) right of way stakes, (b) clear and grabbing stakes, (c) construction limit stakes, (d) vertical reference stakes, (e) cut and fill stakes, and (f) others.

8.1.2 Equipment for Both On- and Off-Road Operations

There is a wide variety of heavy machinery involved in highway construction. Below is a brief introduction to the key types of machinery involved.

8.1.2.1 Bulldozer

A bulldozer is a type of heavy construction machine powered by a tracked tractor and equipped with a heavy metal blade to push soil, sand, or other loose material during construction. It is one of the most common types of construction site preparation equipment. The front metal blade can be changed to different styles to fit the type of work required.

8.1.2.2 Excavator

An excavator is another type of heavy construction equipment. An excavator consists of a boom, a dipper, and a bucket. An excavator's primary function is to dig, which makes it particularly efficient in trenching and in certain earth cutting situations.

8.1.2.3 Loader

A loader is a type of heavy machinery used to either scoop up material and transfer it to other transport vehicles (e.g., dump trucks) or to move material for a short distance from one place to another. Loaders come in a wide range of configurations. The most common ones are bucket loader, front loader, and front-end loader. Because it can also do the basic tasks of a dozer, and an excavator, loaders are one of the most popular and versatile pieces of construction equipment.

8.1.2.4 Motor Grader

A motor grader is used in leveling roadway surfaces with minor cut and fill actions through its forward blade pushing action. It is used for fine grading of a roadway's vertical alignment. Blades used in motor graders are typically long. Modern GPS grade control technology enables a grader blade to automatically follow the design specification (vertical profile), thereby delivering improved precision than what would be manually achievable.

8.1.2.5 Compactor/Roller

A compactor or roller is a type of heavy machinery used to condense soil, gravel, asphalt, and concrete. Compaction is critical in highway construction. There are several types of rollers, including (a) sheepsfoot rollers, (b) pneumatic tyred rollers, (c) smooth wheeled rollers, (d) vibratory rollers, and (e) grid rollers.

Compaction effectiveness is based on machine weight (mass) and roller contact surface area (area) for those not equipped with any vibratory mechanism. The higher the mass/area ratio (pressure), the more effective compaction the machine delivers.

For vibratory rollers, significant extra force is generated by vibrations through rotations of an eccentric shaft inside a roller's drum. The frequency of the vibration and the force from such vibrations are critical to a compactor's overall effectiveness.

8.1.2.6 Pavers

A paver lays asphalt on a prepared base during highway construction. Asphalt is typically placed flat, segment by segment, by a paver. Once the asphalt is laid, it is immediately compacted by a roller. Tracked and wheeled pavers are the most common type used in the industry.

8.1.2.7 Portland Cement Concrete Transport Truck

A Portland cement concrete transport truck is a large commercial truck requiring drivers to possess a commercial driver's license (CDL). Concrete trucks deliver and dispense ready-to-pour Portland concrete to job sites. Concrete trucks can be loaded with either ready-mixed concrete from a mixing plant or dry material (cement and aggregate) plus water. If dry material and water are carried by a concrete transport truck, the mixing is typically done during transport. To maintain the liquid pouring phase of the concrete during transport, agitation or drum turning is deployed until delivery. Inside the drum, a rotating spiral blade is deployed to dispense concrete at a worksite.

8.1.3 Skilled Labor

In addition to operating machinery, basic construction knowledge in the following areas is required from the workforce.

- Aggregate sampling and testing
- Asphalt paving
- Asphalt plant operations
- Concrete field testing
- Concrete lab testing
- Concrete batch plant operating
- Pile driving
- Drilled shaft operations
- Basic field construction estimation

State highway departments prescribe required knowledge and experience for the workforce. Certificates and licenses are used as evidence for meeting such requirements. Training courses in all the subject areas are provided by both private businesses and State highway departments.

The Asphalt Institute and the American Concrete Institute offer extensive training on all the subject areas listed above. Additionally, the Transportation Curriculum Coordination Council also prescribes and provides training programs for workforce development.

8.1.4 Flagger

One of the unique aspects of highway construction is its interaction with traveling vehicles in work zone areas. During highway construction, a complete shutdown of traffic flow is rare. Ensuring worker safety while maintaining traffic flow is challenging. Highway flagging is critical to meeting the challenge (Figure 8.1).

Figure 8.1   A flagger using a traffic paddle to direct traffic flow.

Flaggers direct and control traffic at roadway construction sites. They are responsible for placing all speed limit signs, flags, cones, warning signs, barrels, and barriers per traffic control plans. Flaggers may also need to direct traffic by carrying signs (traffic paddles) or operating pilot cars.

The Manual on Uniform Traffic Control Devices (MUTCD) for Streets and Highways establishes standards for installing both temporary and permanent roadway signs and control devices. Chapter 6E of the MUTCD describes specific needs for flagger control. State transportation departments require trained or certified flaggers. A trained flagger is defined as a person who attends the required training class for flagging but does not take or is not required to take and pass an examination. A certified flagger is a person who has taken the prescribed training and has passed an examination. Private training vendors authorized by State highway agencies offer a wide range of training on flagging. Flagger training typically covers the following areas:

■ Understanding a flagger's responsibility
■ Understanding traffic control zones
■ Understanding proper staging areas
■ Understanding driver behavior and communication needs in dealing with drivers
■ Knowing the proper ways to place traffic control signs and devices
■ Understanding stop, slow, and proceed flagging procedures
■ Knowing how to use traffic paddles
■ Understanding both one-way and two-way closures

8.1.5 Construction Engineer

Senior construction personnel such as construction engineers are responsible for managing both in-house and subcontractors at a job site and acting as a point of contact for construction contract oversight. The person is typically required to

(a) understand contracts, (b) be familiar with construction regulations and codes, (c) possess technical skills and knowledge related to highway (bridge) construction, and (d) be a licensed professional civil engineer in the appropriate jurisdiction (State).

The Federal Highway Administration, through its National Highway Institute, offers a wide variety of training as related to highway construction engineering.

8.2 Construction Contract Administration

A roadway project is most likely a public project meaning a government agency owns the project and is responsible for it. While planning, analysis mandated by the National Environmental Policy Act (NEPA), engineering design, and right of way acquisition are often carried out by governmental employees only, construction is always carried out by private businesses through contracts. Contract administration from both government and private contractor standpoints is critical in successfully delivering a construction project.

State transportation departments are the most common owners of highway construction projects. Duties for state transportation departments for construction projects extend through six major phases: pre-bid, bid advertisement, bid opening, bid awarding, construction engineering and inspection, and final project acceptance and payment.

Steps involved from bid package preparation to bid award are illustrated in Figure 8.2.

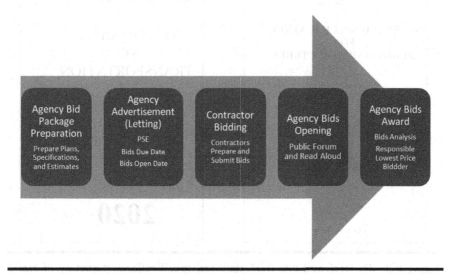

Figure 8.2   Bidding process illustration.

8.2.1 Pre-bid: Bid Package Preparation

A bid package has three key components, design plans (P), specifications (S), and quantity (estimates-E). The entire package is often referred to as the "PSE" package. While design plans and quantity are project-specific (changes from one project to another), a state Department of Transportation (DOT)'s standard specification (S) is typically a stand-alone independent department publication, applicable to all construction projects and contracts.

State DOT standard specifications (S) typically cover and specify:

a. contract proposal forms and format,
b. general contract provisions,
c. general contract terms and agreement,
d. bond requirements,
e. general construction specifications,
f. construction material specification,
g. standard pay items.

A state DOT standard specification is always extensive (typically over a thousand pages). Figure 8.3 illustrates the standard specifications published by the Indiana DOT and the North Carolina DOT. Both government and construction company

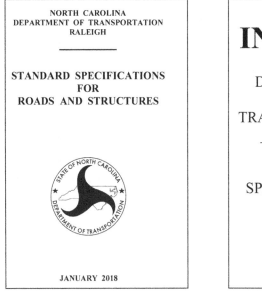

Figure 8.3 IN DOT and NC DOT standard specifications. (Courtesy of North Carolina DOT at www.ncdot.gov and Indiana DOT at www.in.gov/dot.)

personnel need to be familiar with such specifications. Once a contract is executed, a state's standard specification is incorporated as part of a binding contract.

In addition to state transportation department standard specifications, there may be additional forms required. For Federal-aid projects, additional forms as these listed below are normally required as part of the bid package:

- Buy American Act clauses
- Disadvantaged Business Enterprise/Minority Owned Business commitments
- Davis-Bacon commitment (prevailing wages for local workers)
- Nondiscrimination/nonsegregated facilities for civil right protection commitment
- Non-collusion statement protecting the integrity of the highway program from bid-rigging activities

8.2.2 Contractor Prequalification

To ensure an efficient and effective construction process and to protect the public interest, public agencies often adopt various screening processes to determine whether or not a contractor is qualified to bid on a given type of work.

The screening process evaluates a business's work capacity and its workforce experience. Key factors considered include (a) detailed financial statements from the firm minimum covering the previous two to three years, (b) ownership and management control of the firm, (c) equipment and other assets owned by the firm, (d) experience of both individuals and the business, (e) performance – both individuals and the firm as a whole, and (f) current capacity.

While the prequalification process offers an excellent opportunity for the public agency to gain relevant information on a firm, the prequalification process must not limit or prevent competition. The prequalification process must conducive to open, fair, and competitive bidding.

8.2.3 Contractor Bonding Requirement

To protect the public interest, state highway agencies often require construction companies to obtain bonds. A bond is an instrument purchased by a contractor from an insurance company to guarantee that the contractor will perform the agreed-upon work. There are four bond types used in highway construction: bid bond, performance bond, payment bond, and warranty bond.

A bid bond ensures that a bidder will sign the contract if that bidder is the winner of a contract. A performance bond ensures that a contractor will complete the contract. A payment bond ensures that a contractor will pay for the materials, equipment, and labor used on a contract. A warranty bond ensures that work completed will perform as specified in the contract.

While bonds protect public agencies by transferring risk to insurance companies, bonding does cost money and ultimately increases the cost of doing business. Also, new companies may have difficulty in obtaining bonds, which are not conducive to innovation and competition. While bonding, requirements do protect public agencies, they should not be excessive, causing unnecessary project cost increases and preventing competition.

8.2.4 Contract Letting

One of the most important terms used in construction is "let" or "letting." "Let" or "letting" is to make a project available for bidding. Letting date refers to the date a construction contract is advertised. In the advertisement, the nature of the construction work, bid due time, bid due to place, bids open and read aloud date are specified.

8.2.5 Bid Advertisement

Bid advertisement is an announcement to the public and all interested parties that an agency is soliciting bids to perform a described work. Methods used for bid advertisement are governed by state or federal law. Bid advertisement often takes the form of classified advertisement in a newspaper, trade journals, or electronic web advertising.

An advertisement period usually lasts for three weeks. For large and complex projects, offering additional time for contractors ensures responsible bidding. Responsible public agencies may also organize a pre-bid meeting to answer questions from prospective bidders.

The essence of any bidding process is to assure free and open competition. Any requirement that hinders fair competition is to be avoided.

8.2.6 Bid Opening

At the conclusion of the advertisement period per a bid's advertised specification, the public agency opens all sealed bids in a public forum and reads them aloud. The location, date, and time should be consistent with what is specified in the bid advertisement.

Bidders and other interested parties and citizens are present during a bid opening. Such an opening may also be live-streamed and broadcasted on the Internet to enable participation. Public agencies can choose to read aloud each bid item by its cost or total cost.

Occasionally, a bidder's bid package may be considered irregular per a state's standard specification. The most common bid irregularities include (a) failure to sign a bid, (b) failure to provide a required bid bond, (c) failure to offer a unit bid price for each item, and (d) failure to provide or sign a commitment form. A bid that is considered irregular will not be read aloud during the bid opening.

8.2.7 Bid Award

Upon the conclusion of the public opening and reading aloud of all bids, the responsible public agency will conduct a bid analysis to ensure the lowest possible price is received.

Bid analysis includes:

a. Comparison of unit price with agency estimation,
b. Comparison of material quantity,
c. Balancing with regard to cost allocations among material, labor, and schedule,
d. Discovering any significant unit price difference with agency estimates and among all bidders,
e. Deciphering any other discrepancies and irregularities.

Upon completion of bid analysis, the contract with the lowest responsible bid is awarded.

For Federal-aid highway projects, any award or rejection of the lowest bidder requires FHWA's concurrence.

8.3 Construction Engineering and Inspection

CEI refers to all post-bid award activities carried out by the public contracting agency in managing, overseeing, and inspecting work performed by a private construction company.

8.3.1 Post-Award Management

Critical aspects of post-award managing activities include:

- Getting to know all the contractors involved. The prime contractor is the one that executes the contract with the government. The prime contractor may hire another contractor to perform certain tasks per the executed contract. A contractor who is working for the prime contractor is called a subcontractor,
- Progress and schedule monitoring – compare actual field progress vs. the pre-established schedule.
- Cost and schedule monitoring – track payment, progress, and schedule.
- Change of order analysis – carefully consider any change of order, keep track of how often this occurs, how it is resolved, and how long it takes to resolve.

- Differing site conditions – carefully consider such claims, be familiar with the agency standard specification, notice how often such claims are made, how such issues are resolved, and how long it takes to resolve each issue.
- Contractor self-performance – ensure the minimum specified amount of work is truly performed by the prime contractor.
- Subcontracting – ensure the qualification and approval of subcontractors.
- Sampling and testing need for a project – adequate sampling, testing, and record-keeping are carried out.
- Field control – detailed diary covering events, communications, and observations are prepared.
- Inspection reports – ensure completeness and legibility, time, and date.
- Material reports – complete on-site material records by time, date, and utilization.
- Correspondences – ensure complete records-keeping.

The most important aspect of CEI work is the awareness of a contractor's day-to-day operations. Such knowledge enables appropriate decision-making and actions, which can be taken at the earliest stage possible to protect a public agency's interest, and ensures the roadway is constructed per the specification.

Historically CEI functions were carried out by government employees only. Nowadays, private contractors routinely perform CEI duties on behalf of the government. On Federal-aid highway projects, private contractors can only perform CEI duties when a government project engineer provides supervision. The government engineer must have direct responsibility and authority on any Federal-aid project.

Federal oversight and requirements are different for different types of roadways and construction types. Requirements and oversight are more stringent on the National Highway System (NHS) roadways than non-NHS activities.

8.3.2 Change of Contracts

An executed contract is a legally binding agreement between two parties. Contract changes and modifications require both parties' consent. It is common knowledge and experience that not all construction projects can be built exactly per the design plans and specifications.

Regardless of how diligent and thorough design engineers are, there will be unforeseen conditions. When soil is moved during construction, the soil may not be the same as what was originally specified in the design plan. Unknown soil contamination may be discovered, and unknown utility pipes or cables may be present. Additionally, when traffic control is set per a Traffic Control Plan outlined in the construction plan, traffic patterns may change, and the plan may not work anymore. All these issues may result in the need to change the executed contract and specifications to reflect actual field conditions.

Furthermore, a contractor may come up with alternative solutions through their own value engineering analysis, leading to savings for the agency and increased profit for the contractor.

Contract changes may include design plan change, construction specification change, cost change, and schedule change.

8.4 Resolving Construction Dispute

Disputes between a construction contractor and the related public agency are not unusual. Such disputes often arise from (a) project uncertainties (e.g., unforeseen site conditions, design errors and omissions, schedule change, minor cumulative changes leading to schedule issue), (b) process problems (e.g., method of communication, time it takes to make a decision), and (c) people issues (e.g., personal style, management style, communication style, general likeability).

If a dispute is not resolved or resolved quickly, it may manifest into a larger issue jeopardizing a project and putting the public at risk.

Disputes can be litigated in court. However, court litigation is lengthy and expensive. To avoid and minimize litigations, both government and contractor personnel are encouraged to resolve a dispute at a project level through partnering at the earliest time possible. Partnering develops a relationship promoting mutual and beneficial project goals through trust, cooperation, and teamwork. A plan to have open communications and commitments to participate is key to its success.

In addition to practicing partnering, the construction industry has developed various alternatives to resolving disputes without having to go through courts (Figure 8.4).

8.4.1 Negotiation

An issue resolved without any third-party involvement through a dialogue between the contractor and the public agency is called negotiation.

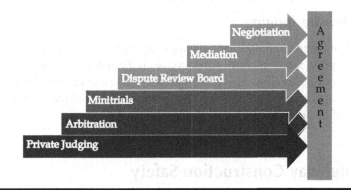

Figure 8.4   Alternative dispute resolution methods.

8.4.2 Mediation

The public agency and the contractor jointly appoint a neutral third party who then facilitates a resolution of the dispute. The mediator does not decide the dispute outcome. The mediator can discuss the issue with each party alone or in a joint session. The resolution of the dispute is still up to the two parties involved through this facilitated dialogue.

8.4.3 Dispute Review Board

A Dispute Review Board (DRB) is a specially created three-member committee where members meet regularly to resolve disputes during a construction project's lifetime. Each party appoints one person, and the third person on the committee is jointly appointed by both parties. A DRB does rule. However, a ruling issued by a DRB is typically nonbinding.

8.4.4 Minitrials

A three-person panel comprised one senior official from each party, and a neutral one agreed to by both the public agency and the contractor is formed. Issues are presented to the panel by experts and lawyers from both sides. A ruling from the panel is nonbinding, but in practice, it is very likely that such a ruling is final. Materials used in minitrials are confidential and cannot be used in court litigation later.

8.4.5 Arbitration

Chosen by both parties, a one- or a three-person panel hears a case and rules on the dispute based on both facts and law. While the ruling is nonbinding, in practice with almost all cases, an arbitration ruling is accepted by both parties.

8.4.6 Private Judging

The final alternative approach in resolving a dispute before filing a case with a court is through private judging. Both parties agree on the appointment of a retired judge to hear the case and allow the retired judge to make a judgment. Again, as in all alternative approaches, a ruling from the private judge is nonbinding but will likely be accepted by both parties.

8.5 Highway Construction Safety

Highway construction safety is paramount to all involved. Highway construction safety is unique as it affects both the traveling public and construction personnel.

8.5.1 Construction Safety and Standard

The Occupation Safety and Health Administration (OSHA) of the Department of Labor provides extensive regulations on all safety issues and areas. Through Title 29 Code of Federal Regulation, OSHA provides highly detailed specifications and standards covering all safety areas including construction safety.

29 CFR Part 1926 titled "Safety and Health Regulations for Construction" prescribes specific construction safety and health standards.

State government agencies such as the State labor department typically adopt the Federal OSHA standards for implementation. Both OSHA and state/local government agencies develop and deliver a wide range of training materials in partnership with the construction industry to educate construction contractor personnel on both regulations and good/standard practices. Private businesses and nonprofit organizations also deliver training to construction contractors and personnel. Construction training ranges from basic awareness and recognition of construction hazards to comprehensive construction safety compliant implementation requirements.

OSHA identifies (a) falls from heights, (b) trench collapse, (c) scaffold collapse, (d) electric shock, (e) failure to use proper personal protective apparatus, (f) repetitive motion injury as the most encountered construction hazard.

8.5.2 Work Zone Safety

For highway construction, work zone safety and mobility are key focus areas for both the FHWA and State highway agencies. Through 23 CFR Part 630 Work Zone Safety and Mobility, the FHWA has prescribed specific organizational-level and project-level procedures to achieve highway construction zone safety for both workers and the traveling public.

The 23 CFR Part 630 calls to:

a. Institutionalize work zone procedures at the organizational level with specific applicable language at the project level.
b. Adopt a system engineering approach, starting with planning and progressing through project design.
c. Address broader impacts of work zones by developing transportation management strategies that address traffic safety and control.
d. Emphasis on partnering driven approach.
e. Develop the ability to adopt relevant provisions per project-specific needs.

The MUTCD offers seven fundamental principles for temporary traffic control covering work zones, as summarized below.

1. General plans should be developed and implemented to provide safety protection to motorists, workers, and other personnel through permanent fixture design specifications by experienced and knowledgeable personnel.

2. Impacts on vehicle travel should be minimized by considering adequate detour and off-peak construction.
3. Clear, positive, necessary, and intent consistent warning signs, delineation markers, and channelization should be set for all roadway users covering all approaching zones and actual work zones.
4. Ensure all temporary traffic control measures and devices operate appropriately as work progresses through daily inspection by personnel who are knowledgeable about the principles of proper temporary traffic control.
5. Ensure roadside safety throughout the temporary traffic control lifetime by providing roadside recovery areas, utilizing crashworthy devices, and offering appropriate storage for nonconstruction vehicles, material, and debris.
6. Ensure that all involved, including upper-level management, receive adequate temporary traffic control training.
7. Proactively establish and maintain good public relations by assessing all roadway user needs, proactively involving news media, and directly notifying abutting property owners and emergency service providers.

For a given temporary traffic control zone, there are four very distinguishable areas involved, the advance warning area, the transition area, the activity area, and the termination area. Figure 8.5 illustrates the basic delineations of such a zone.

1. Advance Warning Area

   An advance warning area refers to a segment of highway installed with signs informing road users about the upcoming work zone. The distance ahead for placing such warning signs depends on the roadway way type and roadway speed. For freeways and expressways, the distance should be longer than that

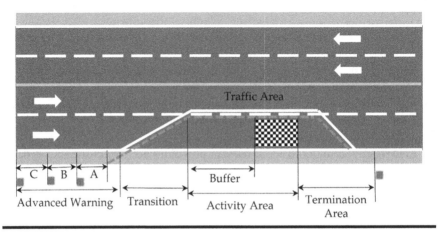

Figure 8.5 Components of a temporary traffic control zone. (Redraw based on FHWA MUTCD Chapter 6 Figure 6C-1, https://mutcd.fhwa.dot.gov/htm/.)

Table 8.1 Recommended Advance Warning Sign Minimum Spacing

	Distance between Signs		
Road Type	A (ft)	B (ft)	C (ft)
Urban (low speed)	100	100	100
Urban (high speed)	350	350	350
Rural	500	500	500
Expressway / Freeway	1,000	1,500	2,640

Source: Courtesy FHWA MUTCD Chapter 6 Table 6C-1, https://mutcd.fhwa.dot.gov/htm/.

for other roadways, because drivers are conditioned to uninterrupted traffic flow. Often this distance extends beyond half a mile. For urban streets, the distance (feet) is typically four to eight times the speed limit in mph.

Table 8.1 provides a general guideline for such distance requirements. The ultimate determination should be based on local engineering judgment.

2. Transition Area

A transition area is a segment of roadway where vehicles are directed out of their normal path before reaching the work activity area. While stationary markings and guide signs are always desirable, mobile devices and signs, such as those mounted-on vehicles, are acceptable too.

3. Activity Work Area

An activity area is a segment of a roadway where construction takes place. It has three separate areas: workspace, buffer space, and traffic space. The workspace is typically designated by channeling devices or by temporary physical barriers. The buffer space is a longitudinal, lateral offset area separating vehicles from the workspace. The remaining lane(s) where traffic passes through is termed traffic space.

4. Termination Area

A termination area is the segment of a roadway where vehicles return to their regular path.

8.5.3 Worker High-Visibility Safety Apparel

All highway construction workers are required to wear high-visibility safety apparel.

The MUTCD Sections 6D and 6E state that all workers, within the right of way exposed to traffic shall wear high-visibility safety apparel that meets the Performance Class 2 or 3 requirements of the ANSI/ISEA 107-2004 "American National Standard for High-Visibility Safety Apparel and Headwear."

While neither the American National Standards Institute (ANSI) nor the International Safety Equipment Association (ISEA) is a governmental regulatory agency, organizations like these represent expertise in the development of safety standards. Governmental agencies often codify such industry standards through their rulemaking authority. In this case, the industry standard for high-visibility safety apparel established by the ANSI/ISEA is codified as the Federal standard by the FHWA through Title 23 of the Code of Federal Regulation via the MUTCD regulation-making authority.

High-visibility safety apparel offers workers conspicuity during both daytime and low-light conditions such as nighttime. This type of clothing often consists of conspicuously colored fluorescent material and conspicuous retroreflective material.

Fluorescent material reemits lights upon adsorbing incoming light with different wavelengths. The phenomenon of fluorescence is often described as "glowing." Retroreflective material is a material reflecting a high percentage of light back (same wavelengths) to or close to its point of origin.

For a given piece of apparel, the standard specification typically covers:

- Surface areas of background material and fluorescent material surface areas. For example, for a jacket, the minimum square inches of fluorescent material required.
- Amount of retroreflective material amount. For example, for a jacket, the minimum square inches of retroreflective material required.
- Width of minimum reflective material. For example, the minimum width of the retroreflective strip required for a jacket's shoulder strap.
- Design and placement of retroreflective material. Patterns on how to place retroreflective material in combination with the fluorescent material.
- Conformity by an accredited laboratory. Before any such apparel can be marketed, compliance with specific standards must be met and certified by an accredited laboratory.

8.6 Next Step

When field construction is complete, the road opens to traffic. Obviously, wear and tear of the road occur right away. To ensure the road performs at an optimum condition and as designed, operational activities such as traffic signal timing adjustment, signage adjustment, deployment of various traffic controls, etc., will become the focus. Also, maintenance will step in, too, ensuring the physical condition of the roadway is adequate and acceptable.

It may take months to years for a construction project to close after a road opens to traffic. The closing delay associated with construction projects is often a result of claims, disputes, and warranty issues.

8.7 Summary

Highway construction is often considered the last step in delivering a transportation project. Private contractors perform virtually all highway construction work. The lowest responsible bidder is awarded the construction project. To ensure the bidding process is fair and the public interest is protected, highway agencies often adopt rigid procedures and selection criteria.

Contract administration is a very critical element for both the project owner and its contractor for a construction project. The success of contract administration ensures that a project's schedule is maintained and its budgeting is not overrun. The construction industry has developed over half a dozen effective approaches to resolve disputes.

Highway construction safety is another element where all construction personnel, including management, need to pay close attention. From simple compliance with worker high-visibility safety apparel needs to advanced construction zone traffic control strategies, following established standards and practices are vital to preventing injuries and avoiding fatalities.

8.8 Discussion

1. State DOTs often implement prequalification requirements for construction companies to qualify for certain activities before they can submit bids under the auspicious of responsible bidding. However, it is also common knowledge that prequalification requirements may limit innovation and prevent fair competition. How a State DOT can balance such competing and conflicting needs effectively.

8.9 Exercises

1. The construction of a highway typically only starts when the final design is completed, the needed right of way is acquired, and all environmental permits are in hand. True or False.
2. Skilled labor and experienced equipment operators are critical in carrying out highway construction. True or False.
3. Skilled labor for highway construction is typically required to be certified or trained for the subject area. True or False.
4. Land surveying is a regulated profession by State agencies, meaning that only duly licensed persons can practice such activities. True or False.
5. Construction survey is to mark various points, elevations, and boundaries in the field per the construction plan to provide visual cues to field personnel. True or False.

6. Match the types of machinery on the left to the type of key activities on the right.

   a. Loader   1) Level loose soil for a large area

   b. Roller   2) Excavate deep drainage trenches

   c. Concrete Truck   3) Clear roughly 2 acres of wooded area for roads

   d. Excavator   4) Compact the soil to a predetermined density

   e. Bulldozer   5) Delivers premixed concrete cement to a job site

   f. Motor Grader   6) Scoop up gravel and excavate a 4-feet trench

7. A flagger directs only construction vehicles on a construction site. True or False.

8. Construction engineers typically do not need to be licensed professional engineers if they understand construction laws and contracts. True or False.

9. Construction contract administration is very demanding because there are many unknown site conditions and factors that may emerge during construction. True or False.

10. Identify the activities listed below which are part of construction contract administration.

    a. Construction bid package preparation

    b. Contractor prequalification evaluation

    c. Determine whether a noise abatement wall is needed.

    d. Bid advertisement

    e. Bid opening

    f. Construction engineering and inspection

    g. Resolve dispute

    h. Determine claim validity

    i. Perform transportation air quality conformity determination

11. A roadway construction bid package consists of three components. Identify them from the list below.

    a. Environmental permit

    b. FHWA's NEPA approval documentation

    c. Roadway Design Plan Set

    d. Estimates of Material Needed

    e. Appropriate MPO's endorsement

    f. State DOT's Standard Specification for Material and Construction

12. A State DOT's Standard Specification for Material and Construction becomes part of the State DOT's executed specific highway construction contract. True or False.

13. A State DOT's Standard Specification for Material and Construction is a very critical document because it has specifications for material, construction methods, and other relevant standards and procedures related to construction. True or False.

14. Why do State DOTs adopt the prequalification process for private businesses to qualify? Select all applicable ones.

a. Protect the public interest
b. Ensure people or businesses working on public projects are competent
c. Protect existing businesses and ensure they have an adequate amount of work
d. Ensure construction quality
e. Prevent schedule delay and budget overrun due inexperience

15. How many types of bonds are typically used for highway construction projects?
16. Duties for state transportation departments for construction projects extend through six major phases: pre-bid, _____, bid opening, _____, construction engineering and inspection, and final project acceptance and payment.
17. What does "letting date" for a construction project mean?
18. A highway construction contract is awarded to the _____ priced and responsible bidder.
    a. highest
    b. lowest
    c. average
19. Construction engineering and inspection (CEI) are work performed by government employees or private business employees on behalf of the government. True or False.
20. Changing of a construction contract is normal. Government contract engineers can modify contract clauses orally. True or False.
21. The most important aspect of CEI work is the awareness of a contractor's day-to-day operations. True or False.
22. To resolve a construction contract dispute, the best approach is to file a case with the appropriate court. True or False.
23. There are many alternatives for resolving construction contract disputes other than court litigation. True or False.
24. The main reason to avoid court litigation for a construction contract dispute is that court litigation is time-consuming and costly. True or False.
25. Highway construction safety consideration covers both construction personnel and the traveling public. True or False.
26. The MUTCD prescribes detailed procedures for construction personnel to follow as related to construction zone setups. True or False.
27. The American National Standards Institute (ANSI) is not a governmental regulatory agency. However, its specification for high-visibility safety apparel is codified as the Federal standard. Is this true? Yes or No.
28. For a temporary traffic control zone, four areas need to be considered – advance warning area, transition area, activity area, and termination area. True or False.
29. The Occupation Safety and Health Administration (OSHA) of the Department of Labor regulates highway construction safety. True or False.

30. Private company employees can perform all CEI activities on behalf of State DOTs as long as they are licensed engineers. True or False

Bibliography

Asphalt Academy, Asphalt Institute, http://www.asphaltinstitute.org/training/.

Concrete Pavement Construction Training, American Concrete Pavement Association, http://acpa.scholarlab.com/.

Construction Program Management and Inspection Guide, August 2004, FHWA Publication No. FHWA-IF-04-013, https://www.fhwa.dot.gov/construction/cpmi04.pdf.

Highway Work Zones and Signs, Signals, and Barricades, Occupational Safety and Health Administration, U.S. Department of Labor, https://www.osha.gov/doc/highway_workzones/index.html.

Performance-Based Contractor Prequalification as an Alternative to Performance Bonds, August 2014, Publication No. FHWA-HRT-14-034, https://www.fhwa.dot.gov/publications/research/infrastructure/14034/14034.pdf.

Work Zone Safety for Drivers, December 2014, Publication No. FHWA-SA-03-012, https://safety.fhwa.dot.gov/wz/resources/fhwasa03012/.

Chapter 9

Operations

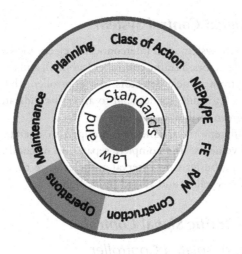

Traffic operations are activities that maintain or increase a road's carrying capacity for both vehicles and people by means other than adding additional through lanes. Activities and approaches include but are not limited to:

- optimizing and synchronizing traffic signals;
- providing more informative highway signs and pavement markings;
- implementing effective access (e.g., median opening, driveway connection) control policy and regulations;
- utilizing intelligent transportation systems such as variable message signs, ramp metering, etc.;
- enabling vehicle speed harmonization;
- taking advantage of modern road weather data and information;
- deploying patrolling vehicles and offering roadside assistance;

- managing travel demand;
- promoting carpooling/vanpooling;
- implementing tolling and other behavior change activities;

9.1 Signal and Optimization

A traffic signal, or traffic light, is a device used to direct vehicle and pedestrian movements at an intersection, ensuring both flow efficiency and safety. After a roadway opens to traffic, traffic signal operations are very closely monitored in the near term, adjusted, and retimed as the flow pattern reaches a stable condition. From there, signals are retimed periodically (e.g., every year) as flow patterns may change significantly.

9.1.1 Traffic Signal Control System

A traffic signal control system is an interconnected electronic system consisting of a number of vehicle detectors, a traffic controller (cabinet), and traffic indicator lights. Information flows from vehicle detectors to a controller. The controller makes a logical decision and transmits its decision to all traffic indicator lights, which are seen by drivers as green, yellow, or red light.

Vehicles on different travel lanes are grouped to operate similarly for their nonconflicting movements. The complete separation of conflicting movements is achieved by adding a buffer time interval and clearance time interval, which are perceived by drivers as yellow and all-red traffic lights.

9.1.2 Types of Traffic Signal Control

9.1.2.1 Pre-timed Isolated Controller

A pre-timed traffic signal controller cycles through a preestablished green-yellow-red sequence with predetermined time durations for the green, yellow, and red indicator lights, regardless of up and downstream traffic conditions or any other traffic controllers.

Vehicle detection sensors are typically not deployed with pre-timed traffic control. The green, yellow, and red lighting sequence repeats itself indefinitely with fixed time intervals.

9.1.2.2 Pre-timed Coordinated Controller

A pre-timed coordinated traffic signal controller cycles through a preestablished green-yellow-red sequence with predetermined time duration for each of the green, yellow, and red indicator light regardless of actual field traffic conditions. However, the starting time of the green light at an intersection has fixed time offsets with its immediate up and downstream intersection green light starting times. For example, three intersections operate under a pre-timed coordinated controller. The middle intersection green light starting time is 38 seconds ahead of its downstream intersection green light and 51 seconds behind its upstream intersection green light starting time. The 38- and 51-second time offsets are predefined and will not change regardless of the traffic condition.

9.1.2.3 Semi-Actuated Traffic Controller

Semi-actuated traffic control is a type of traffic control where traffic on both intersecting roads for an intersection is being monitored. Signal control is partially influenced by what vehicle detection sensors detect.

Semi-actuated control is often used at intersections where a major road intersects with a minor road. The controller allocates green time for the minor road when vehicles are detected, ensuring maximum green time for the main road.

To prevent potential vehicles on the minor road from the excess delay due to vehicle sensor detection failure, a semi-actuated controller typically includes a mandatory green phase for the minor road for a given period.

9.1.2.4 Fully Actuated Traffic Controller

Fully actuated traffic controllers are devices that allocate green time to intersecting roadways based on real-time traffic conditions detected by vehicle sensors on all approaches. While allocations of green time to various movements are dynamic, it also has the preset maximum and minimum green times for various movements.

9.1.2.5 Interconnected Coordinated Traffic Controller

Interconnected Coordinated Traffic Controllers are traffic controllers linked to a central controller where synchronization or coordination of individual traffic controller phasing can be carried out for all vehicle movements. Linked traffic controllers can be both semi-actuated and fully actuated.

9.1.3 Traffic Control Phase Plan

Phase refers to the duration of green indicator light displayed for a specific travel lane or a specific group of travel lanes (lane group), offering protection or permission for vehicles on these lanes to move with the right of way. Vehicles on all other lanes or lane groups are under the red signal indicator light condition. To cover all vehicle movements in all lanes at an intersection, several phases are needed. The time required to run through all phases once is called a traffic controller's cycle length. The cycle length equals to the summation of green phases, yellow phases, and the all read phases in a cycle.

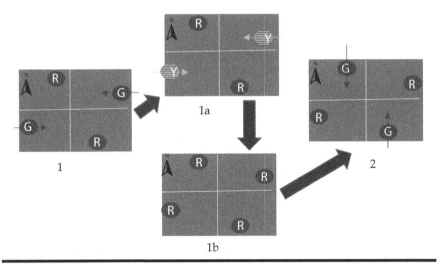

Figure 9.1a   Phase change illustration.

The order of executing various phases controls which lane or lane group vehicles depart first, second, third, and so on, until all lanes have been allowed to depart.

The fewer phases a traffic control cycle has, the more efficient the control plan is. With each phase change, time is lost to an all-red condition (typically ranges from 0.5 to 2 seconds) and efficiency is reduced by the switch as illustrated in Figure 9.1a.

- 1: East-west movements have the right of way – green light. North-south movements are prohibited – red light.
- 1a: Now east-west movements need to clear the intersection – yellow light. North-south movements are still prohibited – red light.
- 1b: East-west movements are prohibited – red light. North-south movements are prohibited – red light. All-red offers additional safety protection.
- 2: North-south movements have the right of way – green light. East-west movement are prohibited – red light.

9.1.3.1 Vehicle Movement

- Protected movements – movements (represented by solid lines in all phasing diagrams) have the right of way as indicated by the green light. Other movements shall yield to these movements.
- Permitted movements – a movement (represented by dashed lines in all phasing diagrams) is allowed only if it is safe. Permitted movement vehicles are required to yield to other movements such as protected movements.
- Pedestrian movements – for compatible travel, pedestrian movements typically have the right of way.

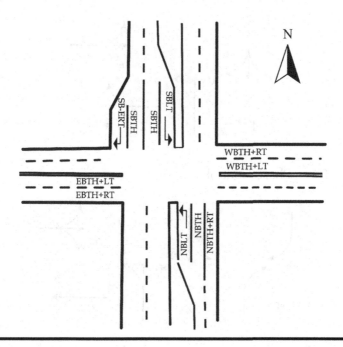

Figure 9.1b Traffic movements at an intersection between a major and a minor roads.

A typical intersection layout with different traffic movements is illustrated in Figure 9.1b.

- NBLT: Northbound left turn (with exclusive left turn bay); NBTH: Northbound through; NBTH+RT: Northbound through + right turn
- SBLT: Southbound left turn (with exclusive left-turn bay); SBTH: Southbound through; SB-ERT: Southbound exclusive right turn (right-turn bay)
- EBTH+ LT: Eastbound through + left turn; EBTH+RT: Eastbound through + right turn
- WBTH+LT: Westbound through + left turn; WBTH+RT: Westbound through + right turn

9.1.3.2 Signal Phasing

9.1.3.2.1 Two-Phase Control

- Phase 1: NBTH and SBTH are under protected green. NBLT and SBLT, NBRT and SBRT are permitted but not protected.
- Phase 2: EBTH and WBTH are under protected green. EBLT, WBLT, EBRT, and WBRT are permitted but not protected.

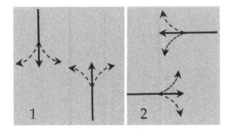

9.1.3.2.2 Three-Phase Control

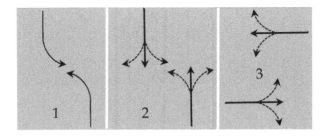

- Phase 1: NBLT and SBLT are under protected green. No other movements are allowed. This type of left protected phasing requires dedicated left-turn bay as illustrated in Figure 9.1b.
- Phase 2: NBLT and SBLT become permitted turn from protected turn. NBT and SBT are protected movements under the green. NBRT and SBRT are permitted turn.
- Phase 3: EBTH and WBTH are protected movements. EBRT, EBLT, WBRT, WBLT are permitted movements.

9.1.3.2.3 Split Two-Phase Control

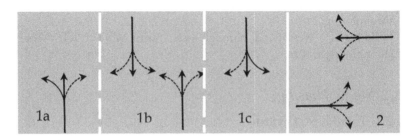

- Phase 1
- Sequence 1a: NBLT and NBTH are under protected green. NBRT is permitted.

- Sequence 1b: NBTH continues to be protected. Now SBTH is also protected. NBLT, NBRT, SBLT, and SBRT are all under permitted condition.
- Sequence 1c: SBTH and SBLT are protected. SBRT is permitted.
- The ideal intersection layout for the north and south approach is to have dedicated left-turn bays. However, under the split two-phase control, lacking of left-turn bays may still work depending on volume.
- Phase 2
- EBTH and WBTH are protected. EBRT, EBLT, WBRT, WBLT are permitted movements

9.1.3.2.4 Four-Phase Control

A four-phase traffic control plan is common for an intersection consisting of two major roadways. Left-turning bays are provided on all four approaches.

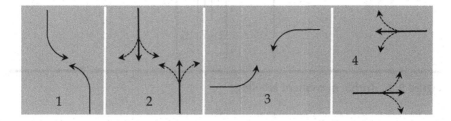

- Phase 1: NBLT and SBLT are under protected green. No other movements are allowed.
- Phase 2: NBLT and SBLT become permitted from protected. NBTH and SBTH are protected movements under the green. NBRT and SBRT are permitted turns.
- Phase 3: EBLT and WBLT are under protected green movements.
- Phase 4: EBLT and WBLT are changed to permitted movements. EBT and WBT are protected movements. EBRT and WBRT are permitted movements

9.1.4 Labeling of Intersection Traffic Movement

By convention, intersection traffic movements are labeled as illustrated in Figure 9.2. They are as follows:

Northbound left turn is movement 1. Clockwise, left-turn movements from other approaches are labeled as 3, 5, and 7.

Southbound through flow is movement 2. Clockwise, through movements from other approaches are labeled as 4, 6, and 8.

Right-turn movements are labeled with two-digit numbers by adding 10 to their compatible through movement label numbers on the same approaches. So, the right turns are 12, 14, 16, and 18 movements.

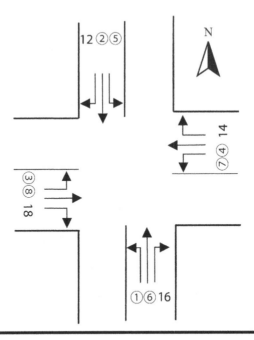

Figure 9.2 Traffic movement labeling.

9.1.5 Traffic Signal Timing Basics

The goal of a signal control plan is to (a) maximize vehicle flow, (b) reduce average delay for all vehicles, (c) minimize extreme delay for low volume movements, and (d) reduce crashes.

9.1.5.1 Effective Green

The green time where movement is maintained at the maximum flow capacity condition in a cycle is called effective green.

9.1.5.2 Critical Lane Group

The lane or a group of lanes where vehicle movements require the longest green time is called the critical lane group.

9.1.5.3 Dilemma Zone

A dilemma zone is a segment of an approaching lane at an intersection where vehicles in the zone can neither stop safely by the stop indicator line nor can they clear the intersection without speeding during a yellow indicator light and before a red

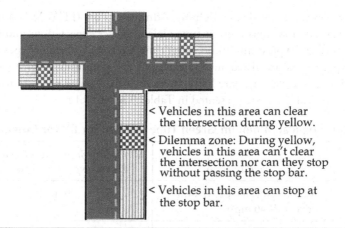

< Vehicles in this area can clear the intersection during yellow.

< Dilemma zone: During yellow, vehicles in this area can't clear the intersection nor can they stop without passing the stop bar.

< Vehicles in this area can stop at the stop bar.

Figure 9.3   Dilemma zone illustration.

signal light turns on (Figure 9.3). To eliminate a dilemma zone, appropriate yellow time should be provided.

The existence of a dilemma zone may cause 1) rear-end crashes when a vehicle tries to avoid running a red light by stopping it in a sudden manner and 2) red-light running in violation of the law, resulting in a potential side or T-bone crash.

9.1.5.4 Adequate Yellow Consideration

The minimum yellow time required to avoid a dilemma zone is the minimum time required for a driver to stop by the stop bar when the driver detects the yellow indicator light. The higher the speed, the longer the minimum yellow time needs to be. The minimum yellow time needed is equal to the sum of a driver's reaction time (the time starting from a driver's discovery of starting of the yellow light to the application of brakes) and braking distance time (the time it takes for a vehicle to stop from when maximum brake force is applied).

The Manual on Uniform Traffic Control Devices (MUTCD) provides detailed guidance on how to determine the exact yellow time and all-red clearance time. Intersection geometric design is the most critical factor for time interval determination. A general rule of thumb is that yellow time should be between 3 and 6 seconds, and the all-red time interval should be from 0.2 to 2.0 seconds.

9.1.5.5 Minimum Green Consideration

The minimum green time is the shortest time interval where a green indicator light is displayed for a given movement. The minimum green time allows a driver to react and act appropriately (driver expectation) to an indicator light change. It should also allow the clearing of intersection approach lane queue and pedestrian crossing.

According to the Federal Highway Administration (FHWA) Traffic Signal Timing Manual, too short a green time violates driver expectations and results in rear-end crashes. Typical minimum green time intervals by facility types to satisfy driver expectations are listed in Table 9.1a. Typical minimum green times for intersection queue clearance are listed in Table 9.1b. Typical pedestrian walk and clearance interval times are presented in Tables 9.1c and 9.1d

Table 9.1a   Typical Minimum Green Time Needed for Driver Expectancy

Phase Type	Facility Type	Minimum Green Needed to Satisfy Driver Expectancy (seconds)
Through	Major Arterial (speed limit exceeds 40 mph)	10 to 15
	Major Arterial (speed limit is 40 mph or less)	7 to 15
	Minor Arterial	4 to 10
	Collector, Local, Driveway	2 to 10
Left Turn	Any	2 to 5

Source: FHWA Traffic Signal Timing Manual, Table 5-3, www.fhwa.dot.gov.

Table 9.1b   Typical Intersection Queue Clearance Time

Distance Between Stop Line and Nearest Upstream Detector(ft)	Minimum Green Needed to Satisfy Queue Clearance[a, b] (seconds)
0 to 25	5
26 to 50	7
51 to 75	9
76 to 100	11
101 to 125	13
126 to 150	15

Source: FHWA Traffic Signal Timing Manual, Table 5-4, www.fhwa.dot.gov.

Notes:
[a] Minimum green values listed apply only to phases that have one or more advance detectors, no stop line detection, and the added initial parameter is not used.
[b] Minimum green needed to satisfy queue clearance, $Gq = 3+2n$ (in seconds), where n = number of vehicles between stop line and nearest upstream detector in one lane. And, $n = Dd / 25$, where Dd = distance between the stop line and the downstream edge of the nearest upstream detector (in feet) and 25 is the average vehicle length (in feet), which could vary by area.

Table 9.1c   Pedestrian Walk Interval Time

Conditions	Walk Interval Duration (seconds)
High pedestrian volume areas (e.g., school, central business district, sports venues, etc.)	10 to 15
Typical pedestrian volume and longer cycle length	7 to 10
Typical pedestrian volume and shorter cycle length	7
Negligible pedestrian volume	4
Conditions where older pedestrians are present	Distance to center of road divided by 3.0 feet per second

Source: FHWA Traffic Signal Timing Manual, Table 5-8, www.fhwa.dot.gov.

Table 9.1d   Pedestrian Walk Clearance Time

Pedestrian Crossing Distance (ft)	Walking Speed (ft/s)		
	3	3.5	4
	Pedestrian Clearance Time (seconds)		
40	13	11	10
60	20	17	15
80	27	23	20
100	33	29	25

Source: FHWA Traffic Signal Timing Manual, Table 5-9, www.fhwa.dot.gov.
Note:
1. Clearance times computed as PCT = D_c / v_p, where D_c = pedestrian crossing distance (in feet) and v_p = pedestrian walking speed (in feet per second).

9.1.5.6 Maximum Green Consideration

The maximum green time is the longest time interval, where a green indicator light is displayed for a given movement in the presence of conflicting demand. The maximum green time is set to avoid excessive delay for the conflicting movement. Example maximum green times under various conditions are listed in Tables 9.1e and 9.1f for general reference.

Table 9.1e   Maximum Green Time for Differnet Roads

Phase	Facility Type	Maximum Green (seconds)
Through	Major Arterial (speed limit exceeds 40 mph)	50 to 70
	Major Arterial (speed limit is 40 mph or less)	40 to 60
	Minor Arterial	30 to 50
	Collector, Local, Driveway	20 to 40
Left Turn	Any	15 to 30

Source: FHWA Traffic Signal Timing Manual, Table 5-5, www.fhwa.dot.gov.
Note 1:
Range is based on the assumption that advance detection is provided for indecision zone protection. If this type of detection is not provided, then the typical maximum green range is 40 to 60 s.

Table 9.1f   Example Maximum Green Time for Different Scenarios

Phase Volume per Lane, veh/hr/ln	Cycle Length (seconds)							
	50	60	70	80	90	100	110	120
	Maximum Green (G_{max}) (seconds)							
100	15	15	15	15	15	15	15	15
200	15	15	15	15	16	18	19	21
300	15	16	19	21	24	26	29	31
400	18	21	24	28	31	34	38	41
500	22	26	30	34	39	43	47	51
600	26	31	36	41	46	51	56	61
700	30	36	42	48	54	59	65	71
800	34	41	48	54	61	68	74	81

9.2 Roadway Signs

Roadway signs provide clear, precise, and timely information and instruction to drivers so that they make timely and appropriate actions (Figure 9.4).

Figure 9.4 Guide and information signs illustration.

The MUTCD provides a set of uniform guidelines for design engineers preparing signage plans. However, until the actual construction is completed and all initial signs are installed, signage plan adequacy is still not field-tested.

Special attention should be paid to reports and complaints received from drivers and the results of an agency's field driving test.

Investigation for potential corrective actions should be carried out to ensure all signs are placed appropriately to achieve their intended purposes.

Roadway sign placement adequacy can also be evaluated through laboratory driving simulators. Laboratory simulation on signage adequacy plays a critical role in the signage adequacy analysis.

9.3 Access Management

Access management is the management of vehicular access and connection points to land parcels situated next to a roadway. During design, access points such as driveway connections and median openings are determined and designed accordingly. However, issues related to access points and median openings do not stop after a road is designed, constructed, and opened to traffic.

Given traffic conditions and land-use change (e.g., a new shopping center is constructed, a new housing complex is built), new driveway access points and median openings are requested by citizens and businesses on a continuing basis. Additionally, existing median openings may no longer be adequate and will need to be closed off or relocated.

Federal, State, and local governments establish laws and regulations in dealing with access management. An effective state DOT's access management program is critical to the safe and efficient operations of all roadways.

9.3.1 Interstate Access (Interchange)

Per Section 111 Title 23 of the United States Code, the Secretary of the U.S. DOT is the only person with the authority to add points of access to or exits from an Interstate highway. State highway agencies can request the Secretary's review and approval on such additions by following FHWA's policy guidance in preparing the needed document known as Interchange Modification Report (IMR) or Interchange Justification Report (IJR).

FHWA's policy guidance is to ensure that a new or revised access point promotes both mobility and safety for the interstate highway system and is compatible and supported by data and information from planning, environment, design, safety, and operations.

There are eight specific policy requirements detailed by the FHWA policy guidance. These eight requirements are as follows:

1. Access Point Needs

    Demonstrate that existing interchanges and local roads in the corridor can't provide the desired access and can't be reasonably improved to accommodate the traffic demand.
2. Reasonable Alternatives

    Demonstrate that operational strategies (e.g., ramp metering, mass transit, and high-occupancy vehicle (HOV) facilities) and other geometric design changes cannot meet the traffic demand.
3. Operational and Safety Analysis

    Demonstrate that the proposed action has no significant adverse impact on the safety and operation of the Interstate facility.
4. Access Connections and Design

    Demonstrate that the proposed interchange connects to a public road, is accessible for all traffic, and meets or exceeds all current design standards.
5. Land Use and Transportation Plans

    Demonstrate that the proposed action is consistent with local and regional land use and transportation plans.
6. Future Interchanges

    Demonstrate that the proposed action is compatible with corridor and comprehensive network operations.
7. Coordination

    Demonstrate that coordination with the area's land development and other transportation system improvements has been carried out.
8. Environmental Processes

    Demonstrate that NEPA coordination and consideration have been carried out.

9.3.2 Access Spacing

For non-interstate and freeway arterial and collector facilities, increasing distance between traffic signals improves a road's capacity, promotes uniform traffic flow, and reduces congestion. State highway agencies typically have state statutes and agency regulations to balance mobility and accessibility.

9.3.3 Driveway Spacing

While driveway connections are necessary, they add friction to traffic flow and reduce a road's traffic carrying capacity. Fewer and farther spaced apart driveways promote safer and more efficient merging, diverging, and flowing of traffic.

State and local statutes typically prescribe specific driveway connectivity rights for abutting property owners (Figure 9.5).

State transportation departments encourage shared driveway connections among adjacent businesses. Establishing shared driveway usage is one of the most challenging issues during preliminary engineering design and the Environmental Impact Evaluation (EIE) phase. Working with affected property owners, residents, and business owners to resolve such matters early promotes safe and efficient roads.

Figure 9.5 Shared vs. individual driveway connections. The left-side photo shows that more than 20 businesses jointly utilize one access point for entry and exit as compared with businesses shown in the right-side photo where each business has its own entry and exit point.

9.3.4 Median Treatments

Medians associated with non-interstate, non-freeway/expressway highways provide dual functions. First, a median acts as a buffer zone separating opposing traffic. Second, a median provides needed space for left-turning or U-turning vehicles. Median treatments, such as median opening, refer to the discontinuation of the continuous barrier and make it available for left or U-turning vehicles without the need for an intersection (Figure 9.6).

9.3.5 Exclusive Left- and Right-Turning Lanes

Dedicated left- and right-turn lanes provide exclusive spaces for both diverging and merging vehicle movements. Such lanes not only offer higher flow capacity but also improve roadway safety.

Figure 9.7 illustrates the two-way left-turn lane (TWLTL) median configuration. This configuration allows continuous left-turn vehicle access without blocking through lane traffic.

Figure 9.6 **Urban median opening for left-turning vehicles illustration.**

Figure 9.7 **Two-way left-turn lane illustration**

9.4 Intelligent Transportation System

Intelligent Transportation System (ITS) is a system of technologies integrated together to improve the operating capabilities of the whole transportation system, covering both capacity and safety. While the ITS terminology may not be well known to the public, the benefits of ITS technology are far-reaching. From the earlier implementation of highway radio, ramp management through metering, dynamic variable message signs, traffic management centers, 511 system, electronic toll collection (ETC), road weather system, to various levels of automated and assisted vehicle operations, ITS has been systematically and continuously improving the entire transportation system. The adoption of fully automated self-driving vehicles in the future will revolutionize how the public travels (Figure 9.8).

From an operations standpoint, the application of ITS takes advantage of the existing infrastructure system by detecting and relaying information to drivers, thereby enabling more efficient and effective ways to travel.

9.5 Traffic Incident Management

On a micro level, a nonrecurring event that occurs on a road causing traffic slow-down and reducing a road's capacity is defined as a traffic incident. Incidents include traffic crashes, disabled vehicles, and cargo spills.

9.5.1 Incident Management Concepts

Traffic incident management is to safely and efficiently clear traffic incidents by aiding the injured, removing the debris, and restoring mobility.

Figure 9.8   Illustration of variable message signs.

An effective traffic incident management system needs to be able to clear an incident promptly by (a) rescuing and transporting injured individuals to hospitals, (b) removing all crash-related debris, and (c) opening the road back to traffic to prevent secondary incidents from prolonged blockage and minimize congestion.

Many parties are involved in traffic incident management. Transportation agencies, law enforcement agencies, fire and rescue, emergency medical service (EMS), and towing and recovery all play a vital and unique role in traffic incident management. Depending on the circumstances, hazardous material personnel covering hazmat and coroner/medical examiner due to fatalities may also be on the scene. Understanding different roles played by different entities is critical to the success of traffic incident management (Figure 9.9).

Transportation agencies' Transportation Management Centers (TMCs), field maintenance staff, and operations road patrols are vital parts of a traffic incident management team. While a TMC serves as the hub for both incident information collection and information dissemination, maintenance personnel and service patrols provide temporary crash site traffic control to ensure the safety of the scene and relay field conditions back to the TMCs.

Emergency 911 dispatchers are typically the first to have incident information on where and what has happened. The 911 dispatchers transmit the information to proper units (e.g., EMS, police, fire, and rescue) to get the appropriate personnel and equipment to the scene.

Law enforcement officers are often the first to arrive at an incident scene. Upon arrival, the on-scene officer assesses the situation, secures the scene, and calls for

Figure 9.9 An incident scene – illustrating various activities. (a) channelizing travel lanes through a crash site, (b) staging recovering and emergency vehicles, (c) crash scene – various emergency vehicles and personnel, (d) further guiding through a patrol vehicle.

additional resources as needed. Additionally, law enforcement officers are responsible for conducting all investigations, including evidence collection.

Fire and rescue personnel, on many occasions, may arrive at an incident scene first. Fire and rescue personnel then secure the scene, provide first aid, and, if warranted, call EMS for further treatment and transport of the injured. Fire and rescue also put out the fire and prevent any potential fire hazard.

EMS provide on-scene triage for the injured and transport to hospitals for care.

Law enforcement or transportation agency personnel notify private towing and recovery businesses to remove disabled vehicles and clear debris.

9.5.2 Road Assistance Patrol Services

One of the most critical elements in incident management is the speed of deploying assistance to motorists. State highway agencies and highway law enforcement routinely provide a wide range of free help to stranded motorists on highways for public benefit.

State transportation departments in partnerships with various other entities such as state highway patrol units deploy incident patrol services along major freeways – especially urban freeways and expressways, providing free assistance to stranded vehicles and vehicles involved in incidents.

Figure 9.10a shows Maryland Coordinated Highways Action Response Team (CHART) patrol vehicles with State Farm (sponsor) decal on the side.

Patrolling vehicles may offer services including:

■ performing minor repairs such as jump start, battery charge, fuel delivery, tire change, and other minor vehicle pull and relocation activities;
■ securing cargo and remove debris;
■ providing first aid to drivers;
■ setting up emergency temporary traffic control; and
■ updating the Traffic Management Center on the current situation and condition.

Figure 9.10b illustrates how various roadways are patrolled by different entities for different time periods in the state of Maryland.

9.5.3 Traffic Incident Management Performance Measurement

Per the FHWA "The Focus States Initiative: On the Road to Success" study, the effectiveness of a traffic incident management program can be measured using three parameters: road clearance time (RCT), incident clearance time (ICT), and the number of secondary crashes.

Figure 9.10a   Illustration of patrol vehicles.

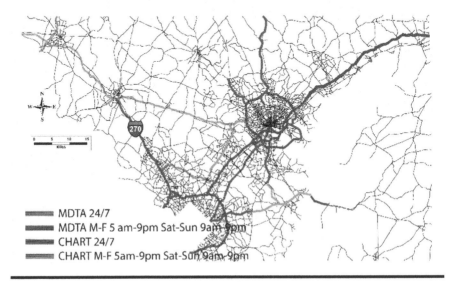

Figure 9.10b   Maryland patroling routes illustration. (Based on CHART at http://www.chart.state.md.us.)

1. Road Clearance Time (RCT)

    The RCT is the time elapsed from when a responsible agency is first aware of an incident to when all roadway lanes are open to traffic.

2. Incident Clearance Time (ICT)

    The ICT is the time elapsed starting when a responsible agency is first aware of an incident until the last responder has left the incident scene.

3. Number of Secondary Crashes

    Within an incident clearance time, the number of additional incidents that occurred within the traffic queues of both travel directions.

9.6 Transportation Demand Management

Transportation demand management (TDM) is the application of methods, strategies, policies, regulations, and laws to increase people's mobility choices to reduce vehicle trips or redistribute trips to alternative periods throughout a day. The ultimate goal of TDM is to reduce congestion.

9.6.1 Operational-Based TDM Strategies

9.6.1.1 Tolling-Based TDM

Historically, tolling was designed to collect money to finance a road's construction, operation, and maintenance. Currently, in addition to its original purpose, tolling is also used as a TDM tool. The practice of tolling as a TDM tool varies widely. Some agencies collect tolls on all travel lanes, tunnels, and bridges at all times with a fixed rate throughout a day. Others collect tolls only on certain lanes with tolling rates varying throughout the day or per the severity of congestion. The practice of tolling varies widely in when and where roadways are tolled.

Tolling used as a demand management tool relies on the market economic theory of supply and demand. The economic rationale is that when the cost is low for using a roadway, demand on the roadway will be high and may exceed available capacity (supply), leading to congestion. The capacity deficiency can be overcome by charging a fee to users, which in turn reduces demand.

9.6.1.2 High-Occupancy Vehicle Lane Strategy

A HOV refers to a motor vehicle with a minimum of two occupants. A HOV lane is a travel lane dedicated to HOV vehicles. State laws establish the criteria on the minimum number of vehicle occupants required for using HOV lanes. Being classified as an HOV vehicle comes with the privilege of being able to operate on dedicated HOV lanes. Regular HOV lanes are free of charge to HOV vehicles. Other vehicles operating on HOV lanes are considered violators and face fines and other consequences (Figure 9.11).

Figure 9.11 HOV lane rules and utilization.

9.6.1.3 High-Occupancy Toll Lane Approach

High-Occupancy Toll (HOT) Lanes are HOV lanes with toll capability. Non-HOV vehicles can operate on HOT lanes by paying tolls while HOV vehicles do not and continue to operate the same manner as any other HOV lanes. This arrangement enables the full utilization of capacity provided by HOT lanes.

9.6.1.4 Variable Tolling Lanes (VTL) Practice

Lanes with variable tolling rates based on the severity of congestion are called variable tolling lanes (VTLs). VTL is designed to offer congestion-free to near congestion-free flow. As congestion sets in, tolling rate increases, which leads to the departure of some vehicles from the toll lane. With fewer vehicles on the VTLs, congestion is decreased.

HOV lane deployment is used to change people's behavior from single-occupancy vehicles to carpooling or vanpooling. Both the HOT lane and VTL operations rely on the theory of behavioral economics associated with supply, demand, and price.

Depending on localities, motorcycles, electric vehicles, and hybrid vehicles may be permitted to operate on HOV facilities even when they are occupied only by a driver alone. The rationale is that these vehicles emit less pollutants than conventional vehicles, thereby leading to better air quality for all.

9.6.1.5 Cordoned Area Tolling

Cordoned area tolling refers to tolling a vehicle when it enters a cordoned geographical area. Such areas are typically heavily congested city centers. Before implementing any cordoned area tolling, a thorough evaluation of the issues listed below should be conducted.

- available public transportation service to the area
- potential business outflow
- potential residence outflow
- net effects on roadway congestion.

9.6.2 Toll Collection

ETC has become standard practice throughout the tolling industry. The biggest advantage of ETC is its ability to charge a toll without the need for a vehicle to stop or slow down. With open road tolls (Figure 9.12a), vehicles can even operate at their normal travel speed.

Figure 9.12a   Open toll through overhead gantry scan antenna.

An ETC system typically consists of four key components: (A) vehicle identification, (B) vehicle classification, (C) toll transaction, and (D) toll enforcement.

9.6.2.1 Vehicle Identification

Vehicle identification determines a vehicle's identity (owner) traveling on a toll road or a toll lane. The most common and reliable method is the usage of radio-frequency identification (RFID) tag (Figure 9.12b) attached to a vehicle. A tollgate antenna communicates with RFID tags on vehicles through dedicated short-range communication (DSRC), gathering needed data.

The RFID tag method has been proven to be highly accurate and reliable. Vehicle operators/owners purchase or lease RFID tags from tolling authorities. The purchased RFID tag is then coded and linked to all similar toll class vehicles planning to use the tag through their license plate numbers. In other words, an RFID tag can be used interchangeably among the same type of vehicles.

Supplementing the RFID tag method, license plate recognition is often deployed to cover vehicles with no RFID tags. The technology relies on processing license plate images captured by cameras installed at toll gates. The automated license plate recognition method enables drivers to use a toll facility without any advance interaction with a toll operator. This method is an effective way of capturing toll violators.

Automatic license plate recognition through image processing still experiences reliability challenges resulting in billing errors and other issues. A manual review is often required to reduce such errors.

9.6.2.2 Vehicle Classification

Toll rates are often different for different vehicle classes. Consequently, proper vehicle classification is critical in assessing correct tolls.

The simplest approach is through prestoring vehicle class information with the RFID tags. However, given that an RFID tag can be moved from one vehicle to

Figure 9.12b   A RFID toll tag.

another, field verification of the actual vehicle class is needed. Field vehicle class verification is accomplished through (a) license plate recognition with data and communication linkages to state motor vehicle registration databases, (b) on-site vehicle axle configuration determination through embedded sensors in the pavement (e.g., loop sensors), and (c) vehicle length and other configuration determination through roadside installed radar or other detection systems.

9.6.2.3 Toll Processing

Toll processing refers to assigning a correct charge to a vehicle owner through appropriate accounting.

Customer accounts can be prepaid or postpaid. With postpaid accounts, customers receive periodic (e.g., monthly) billing for payments. A billing statement typically includes time, date, the facility used, and charges for using the toll facility.

For postpaid accounts and those who have no information registered with a tolling operator, state motor vehicle registration databases (vehicle and vehicle owner) are used to locate vehicle owner names and addresses for billing.

9.6.2.4 Toll Enforcement

Toll enforcement is critical for any successful tolling program. Police patrols at tollgates are effective. However, with an open toll, the on-site patrol is less effective. Instead, automatic license plate number recognition has become a highly effective deterrent to toll evaders because the system can locate a vehicle owner through motor vehicle registration databases.

For the design and operation of a toll collection system, several key issues as listed below need to be thoroughly considered.

- System cost and operational cost – Toll collection cost should be reasonable.
- System compatibility – Avoid designing and operating an isolated and unique system. A functional system shall be compatible and interoperable with other systems in the market.
- Forward-thinking – Given electronic and communication technology evolves at a significantly faster speed than any other highway infrastructure component, a toll system should have the capacity to be adaptable and easily modernized in the near future.

9.7 Traffic Congestion Performance Measures

Traffic congestion performance measures are the collection of data, processing of collected data, and the generation of quantitative parameters to assess how a road or a system of roads functions and meets travelers' expectations.

9.7.1 Average Speed

Average vehicle speed on a roadway or a system of roads during a given period (e.g., 7:30 am to 8:30 am) for a specific date (e.g., July 30, 2019) offers the simplest straightforward information to the public on roadway performance. It is easy to understand and comprehend. However, in the absence of other data such as posted legal speed limit and roadway type, average vehicle speed data alone does not convey a sense of congestion or reliability.

9.7.2 Roadway Level of Service (LOS)

The road Level of Service (LOS) concept is proposed and implemented through the Transportation Research Board's (TRB) Highway Capacity Manual (HCM). Data collected are processed and converted to a lettering system ranging from A to F, where A indicates no congestion, and F represents a roadway under severe capacity constraint operating at a failing and unacceptable grade.

The LOS concept is widely practiced by transportation professionals, public officials, and the public because of its simplistic A–F grading scale. However, the travel condition is a continuous function. To break a continuous function into just six categorical groups is somewhat arbitrary.

9.7.3 Travel Time Index (TTI)

Travel time index (TTI) is a ratio of travel time between the peak travel period and the free-flow (off-peak) travel period. Given peak travel vehicle speed is typically lower than off-peak, travel time for the roadways during peak period is longer than off-peak. The larger the TTI, the slower the peak travel is.

While TTI is very specific, its scale does not convene a direct sense of how slow or fast vehicles move on a roadway.

9.7.4 Average Commute Time

Average commute time data provides travel time from origin to destination regardless of modal choices. It is the best multimodal indicator and offers a true sense of travel experience. The drawback associated with this measure is that such data are challenging to obtain.

9.8 Spot Speed

A spot speed study is a field measurement and analysis of speed at a roadway location during a predetermined time of day under prevailing weather conditions. Prevailing weather condition typically refers to daylight time with good observability and a dry pavement condition. Reasons for a spot speed study may include:

- speed zone needs assessment,
- roadway posted speed limit adequacy analysis,
- citizens' complaint or concerns about speeding vehicles,
- roadway noise complaints or concerns assessment,
- accident analysis.

9.8.1 Measuring Methods

Three methods are commonly used in conducting spot speed measurement. These methods are automatic road tube axle detection, manual radar meter, and manual stopwatch.

9.8.1.1 Automatic Road Tube

The automatic road tube axle detection method is typically applied to multilane and relatively high-volume roads (Figure 9.13a).

Two road tubes are placed on top of the roadway surface parallel to each other but perpendicular to the traveled way. The distance (d) separating the two road tubes is measured and recorded. When a vehicle's front tire contacts the first tube, the sensor connected to the road tube records a signal and the time (t_1) the contact occurred. When the same tire contacts the second tube, another signal is triggered

Figure 9.13a   Road tube layout illustration.

and the time (t_2) when this second event occurred is also recorded. The time difference between t_2 and t_1 is the time the vehicle traveled through the distance d. Travel speed (s) can be computed with a simple formula of:

$$s = \frac{d}{t_2 - t_1}$$

The road tube method is highly reliable. The rubber tube typically lasts more than 48 hours, even under extremely heavy traffic volume.

9.8.1.2 Radar Meter Speed Measurement Method

The radar meter approach is the most popular spot speed measurement method due to its simplicity. The radar meter method can be used in low to medium-volume roadways. Multiple radar meters can be deployed for complete coverage. Radar meters can be handheld or mounted on a tripod. The only requirement for a radar meter is a clear line of sight from the meter to the vehicle. The line-of-sight distance can range from a couple of hundred feet to a couple of miles, offering great flexibility in selecting a monitoring location.

Figure 9.13b shows a sample on how measured speed data can be recorded on a preformatted recoding paper. The pattern of recorded data automatically helps field personnel to determine the pattern of speed distribution.

9.8.1.3 Manual Stopwatch Method

The manual stopwatch method requires no sophisticated instruments other than a stopwatch and a measuring tape. This method can be successfully applied to low-volume roads. The method has virtually no impact on driving behavior and is different from the case of using a radar meter (drivers may artificially slow down due to the presence of radar gun or radar waves) (Figure 9.13c).

Field personnel needs to identify a segment of roadway where an observer from one end of the segment can clearly see the other end of the road segment. The length (d) of the segment needed depends on a roadway's posted speed. The higher the posted speed, the longer the segment should be. In order to ensure good visual detection of a vehicle entering and leaving the zone, segment length rarely exceeds 300 feet. The person holding the stopwatch is stationed at the end of the roadway segment and must be able to visually identify when a vehicle (e.g., front bumper) enters and leaves the zone. The time clocked between entering and leaving the zone is travel time (t). Given the distance (d) is known, speed (s) can be computed by

$$s = \frac{d}{t}.$$

Vehicle Spot Speed Study Form

State of Florida Department of Transportation

Form 750-010-03
TRAFFIC ENGINEERING
10/15

VEHICLE SPOT SPEED STUDY

General Information

Analyst/Observer:	K. Matthews
Agency or Company:	
Date Performed:	6/4/2008
Time Period From:	9:00 AM To: 11:00 AM
Weather/Road Condition:	Dry, Good
Posted Speed (mph):	55 MPH

Site Information

Location:	137 Miles East of Northfork Cr.
City:	
County:	Bedford
Roadway ID:	Douglas Rd. - C.R. 949
Milepost :	28.83
Remarks:	Concealment Quite Difficult

Speed (mph)	Vehicles traveling East bound — Total	Cum Total	Both Directions Total	Both Directions Cum Total
≥ 80				
78 - 79.9				
76 - 77.9				
74 - 75.9				
72 - 73.9				
70 - 71.9				
68 - 69.9				
66 - 67.9				
64 - 65.9				
62 - 63.9				
60 - 61.9				
58 - 59.9				
56 - 57.9				
54 - 55.9				
52 - 53.9	1	104		
50 - 51.9	2	103		
48 - 49.9	2	101		
46 - 47.9	2	99		
44 - 45.9	5	97		
42 - 43.9	9	92		
40 - 41.9	18	83		
38 - 39.9	23	65		
36 - 37.9	21	42		
34 - 35.9	15	21		
32 - 33.9	4	6		
30 - 31.9	1	2		
28 - 29.9		1		
26 - 27.9	1	1		
24 - 25.9				
22 - 23.9				
20 - 21.9				
18 - 19.9				
16 - 17.9				
14 - 15.9				
12 - 13.9				
10 - 11.9				
≤ 10				
TOTALS				

Travel Direction 1 → ← Travel Direction 2

Speed Data Summary	EAST BOUND
85th Percentile Vehicle	88.4
85th Percentile Speed	43.1
10 mph Pace	33 - 43

Both Directions

Figure 9.13b Spot speed data recording paper developed by FDOT. (Florida DOT, Speed Zoning for Highways, Roads and Streets in Florida, Figure 7.1 Vehicle Spot Speed Study Form, www.fdot.gov.)

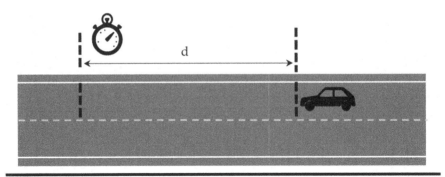

Figure 9.13c   Manual stopwatch method.

9.8.2 Descriptive Statistics of Speed Data

After field data collection, several descriptive statistical parameters are computed.

9.8.2.1 Average Speed

Average speed or mean speed is the arithmetic average of all observed data. It is computed by using the formula of:

$$S_a = \frac{\sum_{i=1}^{N}(S_i)}{N}$$

where S_a is the average speed, N is the number of total measurements, and S_i is the ith individual measured speed.

9.8.2.2 Median Speed

Median speed is a speed residing in the middle of an ordered distribution of all measured speeds. 50% measured speeds are faster than the median speed, and 50% of measured speeds are slower than the speed.

Examples:
18, 19, 23, 24, 25---- median is 23

18, 19, 23, 24, 25, 26---- median is $\frac{23+24}{2} = 23.5$

18, 19, 19, 19, 23, 24, 25---- median is 19

9.8.2.3 Modal Speed

Modal speed is the speed that is observed most frequently among all measured speeds.

Example:
28, 29, 29, 29, 33, 34, 34, 35--- modal speed is 29

9.8.2.4 Standard Deviation

Standard deviation (s) is a statistical parameter measuring the dispersion of the data distribution and can be computed according to the equation below.

$$s = \sqrt[2]{\frac{\sum_{i}^{n}(s_i - s_a)^2}{n-1}}$$

where s is the standard deviation, n is the number of observations, s_i is the ith observed speed, and S_a is the average speed again.

The larger the standard deviation is, the more dispersed the observations are from the mean.

9.8.2.5 Percentile Speed

A percentile is a value on a scale of 0–100, indicating the percentage of a distribution equal to or lower than the value. For example, the 50th percentile speed (median speed) is 41 mph, meaning 50% of all the observed speeds are 41 mph or smaller than 41 mph. The 90th percentile speed is a speed where 90% of observed speeds are lower than it, and 10% of observed speeds are higher. For example, if the 90th percentile speed is 32 mph, then 90% of the observed speed is 32 mph or lower than 32 mph.

85th Percentile Speed $\left(S_{85\text{th}}\right)$ is a speed where 85% of drivers are driving at or below it and 15% of drivers are driving faster than it. The 85th percentile speed has great implications on establishing appropriate posted speed limits for a roadway. The computation of the $S_{85\text{th}}$ can be carried out per the formula below.

$$S_{85\text{th}} = S_a + Z_{85\text{th-one tail}} \times s$$

where $Z_{85\text{th-one tail-normal-distribution}} = 1.036$, S_a is the average speed, and s is the standard deviation computed from all observed speeds.

85th Percentile Computation Illustration

There are 20 measured spot speeds (mph) during a spot speed study listed below. What is the 85th percentile speed?

39, 32, 48, 37, 42, 38, 39, 39, 39, 39, 40, 40, 42, 45, 42, 43, 47, 48, 52, 37

Answer:

$$S_a = \frac{\begin{aligned}(39+32+48+37+42+38+39+39+39+39+40\\+40+42+45+42+43+47+48+52+37)\end{aligned}}{20} = 41.4$$

$$s = \sqrt{\frac{\sum x^2}{n-1} - s_a^2} = 4.7$$

$$S_{85\,th} = S_a + Z_{85\,th\text{-one tail}} \times s$$

$$S_{85\,th} = 41.4 + 1.036 \times 4.7 = 46.3$$

So, the 85th percentile speed is 46.3 mph

9.9 Next Step

Operational activities ensure traffic flow in the desired state. In a parallel fashion, maintenance activities such as mowing grasses, clearing clogged storm drains, replacing damaged guardrails, plowing snow, applying salt during ice condition, etc., further ensure a roadway operable at the desired condition. To a large degree, operational and maintenance activities go side by side, where operations focus on technology and behavior, and maintenance focuses on roadway physical condition.

9.10 Summary

After a road opens to traffic, traffic operational activities begin right away. The purpose of traffic operational activities is to maintain and increase flow capacity and improve safety without the need and expense of adding through lanes and performing other major infrastructure modifications. Activities ranging from access control, signage placements, traffic signal operations, demand management, tolling, to highway patrol services are critical components of such work.

Another significant effort for traffic operations is traffic incident management. Traffic incident management saves lives and maintains traffic flow. The effectiveness of a traffic incident management program is measured in three ways: RCT, ICT, and the number of secondary crashes.

Data gained from operations are used not only for operational decision-making but also for planning to gain a better assessment of overall system improvement needs.

Smart travel apps (e.g., WAZE, Inrix, Google Map) offering real-time congestion information and route choices as developed and implemented by both government and private sectors have proven that the potential from technological innovation is significant. There are ample potential opportunities to reduce congestions and improve safety as new connected, and driverless vehicles are developed and rolled into the markets.

9.11 Discussion

1. Monique drove to see her grandma last Saturday. She said all the traffic lights on her way to grandma's place were operated in a very "stupid" way. She needed to stop at a lot of these intersections because of red lights. She needed to stop even though there were no vehicles at these crossroads. Explain to Monique what might be the cause of her experience.
2. A poor intersection yellow indicator light time allocation may create what is known as a dilemma zone. A vehicle in a dilemma zone faces the impossibility of stopping by the stop bar within the yellow indicator time. The driver may attempt to rush (speed up) through the intersection but will be caught running a red light. Use your words to explain to a traffic court judge why this phenomenon is a possibility and why the "accused" driver should not be held liable for running a red light.
3. Tolling is a very effective tool for managing congestion. However, elected officials and some social groups are often against such measures under the "equity" principle. Discuss the "equity" issue in about 300 words.

9.12 Exercises

1. A traffic signal control system typically consists of a control cabinet, vehicle detectors, and traffic indicator lights. True or False.
2. A traffic signal control system can be an isolated or interconnected system. True or False.
3. An interconnected traffic signal control system always functions more efficiently than isolated controllers. True or False.
4. A pre-timed traffic signal control system has a fixed green, yellow, and red time intervals controlling all movements for an intersection. True or False.
5. A pre-timed traffic signal control system can be fully functional without the deployment of vehicle detectors. True or False.

6. A fully actuated traffic signal control system relies on information detected by vehicle detectors located at all approaching links to allocate appropriate time to each signal phase. True or False.
7. A semi-actuated traffic signal control system is often deployed at an intersection consisting of a minor road and a major road. True or False.
8. A traffic control phase is the period of the yellow traffic light for vehicles in a given lane or a lane group. True or False.
9. A traffic control phase is the period allocated for the green light phase, permitting vehicles in a lane or a lane group to move. True or False.
10. The length of a traffic signal control cycle is the summation of the time deployed for each phase covering a complete cycle. True or False.
11. A vehicle moving under a "protected movement" environment has the right of way. True or False.
12. A vehicle moving under a "permitted movement" environment shall yield to other "protected movement" vehicles. True or False.
13. The more traffic control phases an intersection has, the more efficient the traffic movement for that intersection. True or False.
14. The phase diagram illustrates a _____ (2, 3, 4) phase traffic signal control.

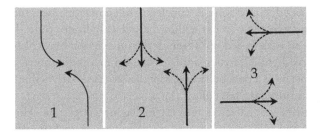

15. Based on the phase diagram illustrated in exercise question #14, it is very certain that left-turning movements from southbound and northbound approaches are heavy. True or False.
16. An approaching lane that has the most vehicles at an intersection is called the critical lane. True or False.
17. A critical lane or lane group requires the longest green time in a signal cycle. True or False.
18. A dilemma zone is created at an approaching leg with an intersection by inappropriate yellow indicator light timing. True or False.
19. The adoption of a minimum green light time length is to ensure that a driver has the time to react and act when faced with indicator light changes. True or False.
20. Typically, the minimum green light time used for intersection traffic control is 10 seconds. True or False.

21. The adoption of a maximum green light time interval is to prevent excess delay for conflicting vehicle movements. True or False.

22. Traffic engineers need to pay special attention to a green phase used to control conflicting traffic movements at an intersection when the green phase timing exceeds 60 seconds. True or False.

23. An intersection signal cycle length rarely exceeds 120 seconds. True or False.

24. Access management for a given road is the control of how abutting properties or other roads are connected to the road. True or False.

25. Driveway connection criteria and median opening standards are unique to roadway functional classes. True or False.

26. The U.S. DOT/FHWA has the final authority in approving new connections to interstate highways or modifications to existing connections to interstate highways. True or False.

27. Traffic incident management involves transportation professionals, law enforcement personnel, medical and emergency service providers, and others. True or False.

28. Traffic incident performance measurements provide quantitative data and information on how quickly and safely an incident is cleared and traffic is restored. True or False.

29. To increase a highway's flow capacity without adding additional travel lanes is defined as capacity improvement work. True or False.

30. Increasing a highway's flow capacity without adding extra travel lanes is defined as operational improvement. True or False.

31. Shifting vehicle flows to different roadways and different times is often achieved by tolling. True or False.

32. Highway tolling not only provides revenue but also reduces traffic demand by either reducing the number of total trips, or shifting trips to a different time, or a combination of the two. True or False.

33. The difference between an HOV lane and a HOT lane is that only high-occupancy vehicles (HOVs) can legally operate in an HOV lane while both HOVs and single-occupied vehicles (SOVs) can travel in a HOT lane. True or False.

34. Traffic congestion performance measures include speed, level of services, travel time index, and other parameters. Any one of these parameters can characterize roadway congestion adequately. True or False.

35. A spot speed measurement is typically carried out for a particular purpose. True or False.

36. The 85th percentile speed is a speed where 85% of the drivers drive faster than it. True or False.

37. The 85th percentile speed is a speed where 85% of the drivers drive slower than or equal to it. True or False.

38. The 85th percentile speed is typically very close to a roadway's posted speed limit. True or False.

39. The 85th percentile speed for a road segment is 45 mph. It is concluded that 85% of all vehicles are traveling at 45 mph or lower. True or False.
40. Three methods are often used for field spot speed measurements. These methods are stopwatch, road tube, and radar. True or False.

Bibliography

Focus States Initiative: Traffic Incident Management Performance Measures, FHWA, https://ops.fhwa.dot.gov/publications/fhwahop10010/fhwahop10010.pdf.

Form Number 750-010-03, Vehicle Spot Speed Study, Florida DOT https://www.fdot.gov/traffic/trafficservices/Studies/MUTS/MUTS.shtm.

Signal Timing Manual, 2nd Edition, NCHRP, http://www.trb.org/OperationsTrafficManagement/Blurbs/173121.aspx.

Traffic Control Signal Needs Studies, Chapter 4C, 2009 Edition, MUTCD, https://mutcd.fhwa.dot.gov/htm/2009/part4/part4c.htm.

What is Access Management? FHWA Office of Operations, https://ops.fhwa.dot.gov/access_mgmt/what_is_accsmgmt.htm.

Chapter 10

Maintenance

Highway maintenance ensures that roads are operable under desired physical conditions and achieve their original design expectations – including their design life expectancies. Key maintenance activities include (a) roadside vegetation management, (b) snow plowing and road surface salt treatment, (c) surface and roadside repairs, (d) debris and garbage removal and drainage cleaning, and (e) other minor none alignment modification work.

10.1 Roadside Vegetation Management

Within a highway's right of way, aside from all traveled lanes, paved shoulders, and sidewalks, the remainder is covered by vegetation established through sodding, seeding, and natural growth. Proper vegetation management is critical for roadway infrastructure health, traveler experience, and aesthetics.

10.1.1 Roles of Roadside Vegetation

Roadside vegetation plays an essential role in the areas listed below.

- Erosion Control

  Aboveground vegetation intercepts rainfall and prevents direct disturbance to the underneath soil from both wind and water. Underground fibrous roots hold soil particles together, preventing them from being separated and carried away by flowing water or blowing wind. This erosion reduction protects the foundation of a road's traveled way.
- Stormwater Management

  Vegetation intercepts precipitation and creates additional time and opportunity for water to infiltrate into the soil below. This reduces stormwater runoff, minimizes stormwater turbidity, and improves receiving water quality.
- Wildlife Habitat Protection

  The proportion of highway right of way covered by vegetation is significant. This is especially true for rural highways. These vegetation-covered areas provide unique habitats for various small animals.
- Aesthetics

  Vegetation within the right of way – especially a well-mowed and edged grassy area, offers an aesthetically pleasing environment to both drivers and the surrounding neighborhoods. Communities and highway agencies often invest a substantial amount of landscaping work in areas where a highway enters or exits a community.
- Noxious and Invasive Weeds Control

  Given the open space of highway right of way and the exposure to travel, the spread of noxious and invasive weeds poses a real threat to native plants. Roadside vegetation management offers an opportunity to eliminate such invasive species.
- Wildfire Control

  It is not uncommon for vegetation to catch fire during a dry season. These fires often create significant safety challenges in areas of both fire hazard and smoke-related visibility. Proper clearing and maintenance of highway vegetation help to eliminate and reduce fire hazards.

Roadside vegetation management with State highway agencies is often codified by state statutes. An integrated management system incorporating vegetation selection, mechanical mowing, animal grazing, and herbicide application is the standard practice.

10.1.2 Mowing

As grasses, weeds, and other plants grow tall, they may block a driver's line of sight, creating a safety hazard. Tall vegetations may also hide animals such as deer from drivers. Collisions between vehicles and deer or other large animals pose life-threating risks to drivers and vehicle occupants. Mowing helps to eliminate and reduce such hazards.

THE MOST CRITICAL ROLE OF ROADSIDE VEGETATION

- Preventing soil erosion by both wind and water.
- Ensuring unobstructed driver line of sight.

10.1.2.1 Mowing Frequency

Depending on local climate and growing season length, mowing may be performed on a monthly or quarterly basis.

A typical guideline is that vegetation height must not interfere with a driver's line of sight and does not affect the sight view necessary for a driver to operate his or her vehicle safely (Figure 10.1a).

State highway agencies typically specify that mowing should be performed when vegetation reaches a height of 12–14 inches. To ensure the regrowth of all the vegetation, mowing height is kept just a bit of higher than 6 inches for all the vegetation.

Figure 10.1a   Median mowing.

Figure 10.1b Roadside wildflower area illustration.

10.1.2.2 Type of Mowing

Normal mowing is mowing without any special restrictions with regard to location, time, and season.

Special mowing is mowing under a set of specific instructions and conditions, such as only after seeds have matured or hatching season is over. Special mowing is most often conducted for areas established by highway agencies or communities for a particular program. For example, areas established for roadside wildflower or special habitat zones are locations that require special mowing.

In the case of wildflower areas, mowing is only conducted after the growing season is over. Mowing helps to reestablish and regrow the flowers for the following year. Certain vegetation zones may be mowed only after the season is over, ensuring ground-nesting bird hatching and other wildlife usages of the areas are over (Figure 10.1b).

10.1.2.3 Mowing Safety

Mowing poses safety challenges to both mowing personnel and the traveling public through two distinct means. Mowing equipment and personnel present attention-grabbing objects for drivers on the road. The distraction creates a high potential for

	A	B	C	D	E	F	G	H	J	K	L
1	30	0.5	0.75	5 D	0.875	2.875	11.908	12.407	10.507	11.007	1.875
2	36	0.625	0.875	6 D	1.25	3.25	14.339	14.839	12.658	13.159	2.25
3	48	0.75	1.25	8 D	1.5	4.5	18.952	19.952	16.711	17.711	3

COLORS: LEGEND, BORDER — BLACK
BACKGROUND— ORANGE (RETROREFLECTIVE)
1: Minimum 2: Conventional Roadways 3: Freeway and Expressways

Figure 10.1c   Mowing warning sign according to MUTCD. (MUTCD Chapter 6G - Figure 10.1.2c, https://mutcd.fhwa.dot.gov/.)

collisions involving vehicles, mowers, roadway structures, and other roadway elements. The Manual on Uniform Traffic Control Devices (MUTCD) Chapter 6G prescribes the standard warning signs (Figure 10.1c) for displaying at locations where mowing is conducted. It is critical for mowing personnel to receive adequate safety training and that mowing personnel abides by all established safety procedures.

The second issue is the mowing activity itself. During mowing, a mower's rotary blades can kick objects (e.g., stones, rocks, wood) out to travel lanes.

These objects may hit traveling vehicles. Given the speed of traveling vehicles, a collision with a projectile can be disastrous. To reduce and minimize objects kicked out by the rotary blades, trash and litter removal from the right of way before mowing is exercised. Also, mowing against the traveling direction of the nearest lane may be practiced. This mowing direction arrangement may result in some of the mower kicked-up objects to travel in the same direction as the traveling vehicles, reducing impacting forces through reduced speed differentials.

SAFETY FOR BOTH MOTORISTS AND MOWING PERSONNEL IS THE NUMBER ONE CONCERN FOR ALL HIGHWAY MOWING ACTIVITIES.

- Appropriate training is a must for all involved.
- For contracting mowing, appropriate contracting terms and conditions must be developed and included to clearly cover all safety issues.

10.1.3 Herbicide Application

In addition to mowing, herbicide application is another effective means for roadside vegetation control. The application of herbicide should be integrated with mechanical mowing to achieve desired results (Figure 10.2).

10.1.3.1 Nonselective Herbicides

A nonselective herbicide affects all vegetation. Such chemicals are typically applied to either weeds or bare ground. The net result is that the entire area is cleared of vegetation for some time.

10.1.3.2 Selective Herbicides

A selective herbicide is a chemical that only affects certain vegetation types (one type of weeds vs. another). It is used to control or remove specific weeds. The key to its successful usage is the correct identification of undesired weeds. The selective herbicide is also applied to either the foliage or bare ground.

Herbicide application is often reserved for areas where mowing is not feasible and areas where noxious and invasive weeds exist.

10.2 Winter Snow and Ice Control

Winter offers a very different environment for both highway travel and highway maintenance. These differences range from sunrise and sunset timing changes,

Figure 10.2   Failed vegetation control associated with urban roads.

low-temperature conditions, freezing rain, and the presence of snow and ice. Keeping roads operable at an acceptable condition is a significant challenge to highway agencies during the winter season.

10.2.1 Goal of Snow and Ice Control

When precipitation falls on a roadway and the temperature is lower than 32°F, the precipitation undergoes a state of phase change from liquid to solid. Ice forms and bonds to roadway surfaces. Ice-bonded pavement surface reduces a roadway's skid resistance (friction between the pavement and tire), creating a dangerous travel condition for all vehicles.

Snow not only covers up roadway delineation markings but also acts as the water source for ice formation. Consequently, snow plowing not only removes the snow but also cuts off the source for ice formation.

Winter snow and ice control achieves four objectives:

■ Ensuring roadway markings and delineations are visible to drivers
■ Preventing ice formation
■ Preventing ice from bonding to roadway surfaces
■ Removing any ice and ice-snow mixtures from roadway surfaces

10.2.2 Mechanism for Snow and Ice Control

10.2.2.1 Snow Removal

Snow removal is typically achieved through mechanical means such as snow plowing. Snow plowing is carried out by equipping dump trucks with plows (e.g., one-way front plows, deformable moldboard plows, side wings, and slush removal plows). Plow width typically ranges from 9 to 12 feet.

Plow trucks may operate in parallel on multilane highways in a closely staggered formation. Such an arrangement enables the placement of displaced snow directly to the roadside.

10.2.2.2 Preventing Ice Formation

It is well known that virtually all salts are freezing depressants. Their presence in water can lower the solution's freezing point (eutectic temperature) further below 32°F as with pure water. The higher the salt concentration is, the further the freezing point is lowered. This characteristic of salt is utilized to prevent ice formation on roads.

EUTECTIC TEMPERATURE (ET)

ET is the highest possible freezing temperature of any solution or the lowest possible melting temperature of any solid mixture.

Ice formation prevention or anti-icing formation requires that salt is applied to a roadway surface at a precise time and temperature. If the salt is applied too early, it is then blown away by traveling vehicles and wind. Additionally, for the applied salt to function, there must be sufficient moisture on the road surface to melt the salt and distribute the salt solution to all pavement surfaces. However, such moisture cannot be excessive because the salt will be washed away.

With the advancement of road weather data and information, preventing ice formation can now be achieved more effectively than ever before.

10.2.2.3 Deicing or Preventing Ice Bonding to Roadway Surface

Deicing is the removal of ice from pavement surfaces and the prevention of ice from bonding to roadway surfaces. The approach is to break the bond between the pavement surface and ice.

Solid salt is commonly applied after a minimum initial accumulation of snow on roadways (half an inch to an inch). It is then spread by traveling vehicles as they mix the salt with snow. The salt melts the snow or ice and forms a solution that prevents further ice formation by depressing the freezing point.

WAYS OF CONTROLLING ICE ON ROADS

- Preventing ice formation
- Breaking bonds between ice and pavement surface.

Table 10.1 Selected Depressant Properties of Various Salts

Chemical	Eutectic Temperature °C (°F)	Eutectic Concentration (%)
Calcium chloride (CaCl$_2$)	−51(−60)	29.8
Sodium chloride (NaCl)	−21(−5.8)	23.3
Magnesium chloride (MgCl$_2$)	−22(−28)	21.6
Calcium magnesium acetate (CMAc)	−27.5(−17.5)	32.5
Potassium acetate (KAc)	−60(−76)	49

Sources: FWHA, Manual of Practice for an Effective Anti-icing Program, Appendix A – selected chemicals and their properties.

Table 10.1 provides the eutectic temperatures for various salt solutions. Comparing with the 32°F temperature for pure water, it is obvious that salt's ability to drive freezing points further down is significant.

10.2.3 Road Weather

Road weather refers to data and information associated with current, near-term, mid-term, and long-term weather conditions for an area about when precipitation will occur, how much precipitation is anticipated, air and pavement temperatures, wind speed and direction, and humidity levels (Figure 10.3).

Local weather forecasting and actual field observations enable the pre-positioning of snow plow trucks in strategic locations. These pre-positioned snow plow trucks are loaded with appropriate material (e.g., salt, skid material, repair material) ready for deployment. When actual temperature and precipitation reach a prespecified condition, salt and brine water are applied to roadway surfaces.

Road weather data are typically collected through a system of connected roadway weather stations installed along roadsides. A station consists of atmospheric sensors (mounted on a tower) measuring atmospheric pressure, wind direction, wind speed, air temperature, humidity, and pavement sensors (embedded in the pavement) to monitor pavement temperature. Weather stations are often linked to a central data gathering system with built-in communication capability.

10.2.4 Road Salt

Road salt used for deicing and roadway treatment is generally mined rock salt with sodium chloride (NaCl) as its main component. Other components

ROAD SALT COMPONENT

Mainly sodium chloride (NaCl) + minor calcium chloride (CaCl$_2$) and magnesium chloride (MgCl$_2$).

Figure 10.3   A road weather and traffic monitoring station.

of rock salt include calcium chloride ($CaCl_2$) and magnesium chloride ($MgCl_2$). Occasionally calcium magnesium acetate ($CaMg_2(CH_3COO)_6$), an industry synthesized compound, is also used. Compared with rock salt, calcium magnesium acetate has a much higher solubility and stronger freezing depressant effect. However, calcium magnesium acetate is much more expensive than regular road salt.

10.2.5 Other Effects from Road Salt

Through its depressant effect on freezing point and its anti-ice-bonding effect, road salt enables the safe operation of highways during snow and ice conditions. However, road salt is also a principal corrosive agent for both vehicles and roadway steel structures. The application of large amounts of roadway salt may also affect the stormwater quality and impact exposed vegetation.

10.3 Pavement Maintenance

Once a road opens to traffic, its pavement starts to deteriorate right away as a result of traffic flows (loading), environmental impacts (e.g., rain, snow, freeze, and thaw), and other factors (e.g., settlement and aging). Pavement undergoes constant change as time goes by. To maintain an optimal pavement condition, highway agencies rely on a systematic approach in their maintenance decision-making. The approach often analyzes and evaluates both individual road and all roads on a system level with regard to investing limited resources and achieves an overall adequate service to the traveling public.

To facilitate such analysis and decision-making, pavement condition data such as (a) pavement surface distress, (b) surface smoothness or roughness, (c) structural capacity, and (d) surface friction are collected.

Pavement surface distress includes cracks, rutting, faulting, potholes, and other deterioration. Surface distress is often detected through optical scans such as human eyes and computerized image processing.

Surface smoothness or roughness is a measurement of pavement surface elevation changes. It is a quantitative characterization of how drivers sense their vehicle ride quality. Pavement smoothness or roughness measurement is carried out through laser scans of pavement surfaces with cumulative vertical distance change per defined roadway length. For example, cumulative vertical distance change for 1,000 feet of roadway segment is measured to be 6 inches. One of the most widely used surface roughness indexes is the International Roughness Index (IRI). The IRI is computed with the accumulated suspension motion (vertically) distance divided by the distance (horizontally) a vehicle traveled where such vertical accumulation has occurred.

Pavement structural capacity is pavement's ability to distribute loads. Structural data can be obtained through both destructive and nondestructive methods. The destructive method cores pavement and analyzes the cored sample. Nondestructive methods can be carried out through the Falling Weight Deflectometer method or penetration radar approaches.

Surface friction property refers to friction coefficient data at the interface between the tire and pavement. The friction affects a vehicle's stopping sight distance, which has significant relevance to safety. Surface friction coefficients can be measured with a variety of methods, including the most common locked-wheel approach.

10.3.1 Pavement Management System

A pavement management system is a decision-making tool where both project (e.g., a specific road) and system (e.g., a group of roads) level decisions can be made regarding when, what action, and how much investment should be delivered to a road or a group of roads ensuring an optimum pavement condition. In other words, a pavement management system facilitates the optimum distribution of funds to different roadways to achieve the optimum pavement condition for the entire roadway system.

A pavement management system consists of:

- mathematical and statistical models on pavement deterioration rates as related to traffic and other factors;
- benefit and cost modeling submodules;
- input data such as current pavement condition data, traffic flow information, traffic flow projections, and desired or minimum acceptable pavement condition;
- output information (e.g., investment needs at various future times, roughness index, surface rutting, present serviceability rating).

Pavement condition improvement is achieved through three main activities, pavement reconstruction, pavement rehabilitation, and pavement maintenance (Figure 10.4).

10.3.1.1 Pavement Reconstruction

Pavement reconstruction refers to actions to replace or reestablish the entire pavement. It involves the replacement of all existing pavement layers. The structural capacity of a reconstructed pavement may be equal to or greater than the original pavement.

10.3.1.2 Pavement Rehabilitation

The purpose of pavement rehabilitation is to extend the functional life of the current pavement. It involves only the surface layer. Resurfacing, Restoration, and Rehabilitation, commonly referred to as 3R work, are classics of pavement rehabilitation activities.

10.3.1.3 Pavement Maintenance or Corrective Maintenance

Pavement maintenance or corrective maintenance is work performed to correct a deficiency such as filling potholes, sealing cracks, grinding faulting slabs, resealing surface, and other actions without increasing or change of a pavement's structural capacity. These activities are aimed at reestablishing a pavement's functionality, maintaining its current overall condition, and slowing down what otherwise would be accelerating deterioration.

Figure 10.4 Milled surface for resurfacing.

10.3.2 Flexible Pavement Failures

There are many types of flexible pavement failures. These failures are often the result of a combination of improper design, low-quality material used, inexperienced construction personnel, inadequate construction method, excessive traffic loading, and adverse environmental conditions (water penetration, extreme temperature, freeze-thaw cycle). Flexible pavement failures are grouped into four categories, as discussed below.

10.3.2.1 Flexible Pavement Cracking

Cracking is pavement breakdown in various patterns or forms such as the checkered alligator crack pattern, block format, along with the edge of traveled way following the longitudinal direction of travel, reflective the underlying layer cracking pattern, or perpendicular to (transverse) the direction of travel.

Alligator cracking also is known as fatigue cracking. It refers to an interlaced pavement crack pattern resembling an alligator's checkered skin.

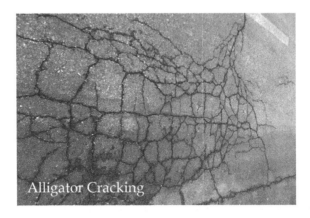

Alligator Cracking

It is a result of differential pavement layer movements under repeated heavy axle loads and is exacerbated by moisture issues associated with base layers and other layers.

Block cracking, as its name implies, denotes interconnected cracks occurring within an isolated rectangular block zone. Block cracks do not have excess longitude or transverse lengths. Block cracking may also occur in non-travel lane areas.

Blocked Cracking

Causes of block cracking are often the aging of the binder material or the usage of inadequate and poor-quality binder material. Binder material's inability to expand and shrink with the daily temperature cycle results in such breakages.

Edge cracking is the formation of longitudinal cracks within a pavement's edge area. Edge cracking is almost always caused by a lack of lateral support from shoulders, inadequate base material or subgrade (settlement), and/or water penetration.

Longitudinal cracking denotes cracks formed parallel with a road's travel direction and over a relatively long distance.

Longitudinal Cracking

Longitudinal cracking is seldom due to traffic loading. The cause is often a result of sliding of roadway side slope and possible differential settlement of filling material.

Transverse cracking denotes pavement cracks that occurred in a perpendicular direction to a road's traveling direction.

Transverse Cracking

Like longitudinal cracking, transverse cracking is not load-related either. Shrinkage of material is the main cause of this type of cracking.

Reflective cracking is cracking resembling the original underlying pavement cracking patterns. Reflective cracking is caused by the underlying pavement layer failures, including settlement and lateral movement.

10.3.2.2 Pavement Deformation

Pavement deformation is a permanent deviation of asphalt pavement from its original state due to repeated vehicle loadings. Pavement deformation mainly exists in forms of corrugation, shoving, and rutting.

Corrugation deformation is a wave-like transverse deformity in the direction of travel. The phenomenon is the movement of surface pavement.

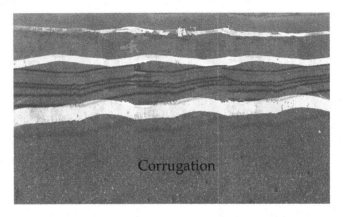

It is caused by an unstable surface layer under a stop-and-go traffic loading pattern.

Shoving refers to localized longitudinal pavement relocation due to layer slippage. It usually occurs where vehicles accelerate and stop. The acceleration and deceleration forces exerted to the pavement structure magnify the underlying issue.

Unstable pavement layer problems are chiefly caused by improper pavement material mix, low material quality, and aging.

Rutting is a permanent vertical deformation of the pavement surface layer or the lateral displacement of pavement material.

Rutting

Rutting presents itself as a depression in the direction of travel along wheel paths. Insufficient compaction during construction and improper pavement mix (e.g., too much asphalt) are the main causes of rutting.

10.3.2.3 Material Failure

Material failure refers to nonperformance of pavement material. This type of failure includes bleeding and raveling.

Bleeding refers to the seepage of bituminous binder used in asphalt pavement mix to the pavement surface, creating a sleek, sticky, and shiny reflective surface. This is a result of too much bituminous binder used in the original asphalt mix or excessive use of tack coat or prime coat during the construction of various layers.

Raveling is the separation and loss of aggregate particles from a pavement surface.

This is often a result of the usage of poor bituminous and aggregate mixture and excess hardening of the bituminous material. The presence of water and heavy traffic accelerate aggregate dislodging.

10.3.2.4 Degradation

Degradation failure refers to problems caused by pavement slippage, potholes, and polished aggregate issues.

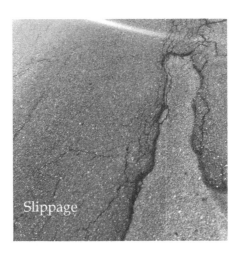

Slippage refers to a clear separation of pavement surface layer from the layer below, covering a relatively large area. Slippage failure is a result of poor bonding between layers and inappropriate application of tack or prime coat.

Potholes are results of the degradation of binding between the surface course and the underlying layer and the breakup of some portion of surface course materials.

Polished aggregate refers to the wearing away of surface binding material resulting in the exposure of coarse aggregate directly to vehicle tires.

10.3.3 Flexible Pavement Pothole Repairs, Crack Sealing and Filling

Pothole repair, crack sealing and filling associated with asphalt pavement are the most common maintenance activities. The activity typically peaks right after the winter season is nearly over, and daily freezing and thawing cycle starts to repeat itself. During the freeze-thaw cycle, liquid water penetrates the pavement through cracks. When freezing, temperatures set in, liquid water undergoes a phase change to solid ice. The volume of the water due to phase change expands below, forcing pavement adjacent to the crack to rise. This "rise" reduces its load-bearing capacity. As this cycle repeats itself, loads from traffic eventually break the weakened asphalt piece, forming a pothole.

While the freeze-thaw cycle exacerbates the formation of potholes, liquid water alone can lead to pothole formation. Seeped in water erodes pavement base material, which makes the base weak. When the base is weakened, the pavement is easier to be broken from load exerted by traffic.

FHWA "Materials and Procedures for Repair of Potholes in Asphalt-Surfaced Pavements-Manual of Practice" and "Materials and Procedures for Sealing and Filling Cracks in Asphalt-Surfaced Pavements-Manual of Practice" prescribe very detailed procedures on fixing such problems.

10.3.4 Rigid Pavement Failure

Asphalt pavement relies on interlocked gravel bonded together through bituminous cement to distribute all traffic load. Its bearing capacity relies on all layers, including the subgrade. Load distribution originating from the tire-pavement interface propagates throughout all layers. On the other hand, rigid pavement due to its rigidity (high modulus of elasticity) distributes its traffic load to a large area, reducing pressure and stress to its immediate underlying layers. This loading distribution and load propagation differences result in different pavement deterioration patterns.

Rigid pavement failure includes linear cracking, durability cracking, shrinkage cracks, warping cracks, corner breaks, scaling, spalling, pumping, faulting, polished aggregate, and others.

10.3.4.1 Linear Cracking

Linear cracking refers to straight-line cracks splitting an individual slab into several large pieces. These cracks can be in any direction (e.g., longitudinal, transverse, diagonal).

Excess traffic load is often the lead cause of such cracks. Such cracks are always exacerbated by adverse conditions such as thermal curling stress (a temperature difference driven expansion differential), inadequate base support, or base moisture penetration.

10.3.4.2 Durability (D) Cracking

D cracking refers to a series of crescent-shaped cracks closely spaced together near a longitudinal and transverse joint corner. It is caused by moisture-absorbing aggregates undergoing repeated freeze-thaw cycles. The destruction starts at the base of a slab with large-sized aggregate particles and typically can't be detected in its early stage.

10.3.4.3 Shrinkage Crack

Shrinkage cracking is intrinsic to the poured Portland cement concrete. These cracks take the form of hairline cracks with a pattern of discontinuity and meandering. Often a crack forms in the concrete, tapers to a stop, and then a new parallel cracking line to the first crack line forms and propagates through the pavement.

Shrinkage cracks occur during setting and curing. These cracks may be longitudinal as well as transversal. Poor reinforcing steel design, contraction joints sawed too late, and improper curing techniques (e.g., allowed to be dried too quickly) further exacerbate the issue.

10.3.4.4 Warping Crack

Warping cracks are cracks near pavement edges and joints. Concrete slabs expand under high-temperature conditions resulting in the development of stress.

The expansion stress eventually breaks the slab near the edge of a joint. Proper longitudinal and transverse joint area reinforcement reduces such cracks.

10.3.4.5 Corner Breaks

Corner cracks occur next to the corner of a pavement slab.

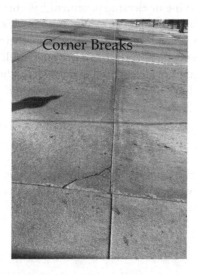

Excess load, lack of proper base, curling, and warping stress all contribute to corner breaks.

10.3.4.6 Scaling

Scaling is the disintegration, flaking, or loss of local pavement slab wearing surface.

Causes of scaling include the usage of non-air-entrained concrete (air entrainment protects concrete during the freeze-thaw cycle), deicing chemicals used, and improper concrete finishing technique.

10.3.4.7 Pumping

Pumping refers to emptying or ejecting pavement base material from underneath a slab due to a change in water pressure. When a vehicle's tire is in contact with the pavement, it exerts pressure on the pavement. If there is water accumulated under a slab, the water will be pressurized. Water under pressure seeks relief paths. Cracks and joints provide such relieving routes. Pumping leads to corner cracks and faulting. A high underground water table and poor joint sealing lead to pumping.

10.3.4.8 Faulting

Faulting refers to an elevation difference between two adjacent slabs across a joint or crack. When the elevation difference is excessive, diamond grinding is used to grind down the elevated part.

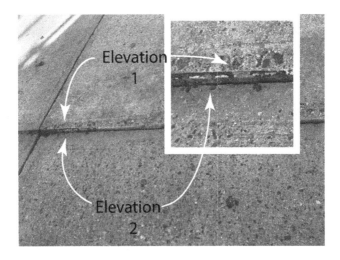

The cause of faulting can be slab pumping, slab settlement, slab curling, and warping.

10.3.4.9 Spalling

Spalling is the cracking, breaking away, or chipping of concrete pavement at joint or crack edges. It leads to loosed debris on highways.

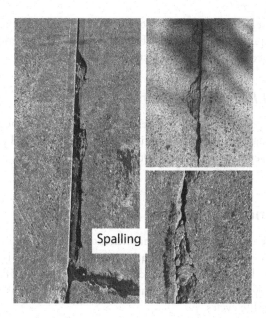

There are several potential causes for spalling, including excess traffic loading, excessive stresses at joints/cracks, freeze-thaw cycle, low-quality Portland concrete cement, or improperly inserted dowels in joints.

10.3.4.10 Polished Aggregate

Polished aggregate refers to the phenomena where the surface binding mortar is worn away, and coarse aggregates are exposed and extended above the mortar cement paste.

Polished aggregate phenomena can occur fast if the aggregate used is susceptible to abrasion. The polished aggregate problem leads to reduced friction at the tire and pavement interface, impacting vehicle stopping sight distance.

10.3.5 Rigid Pavement Repair

As with flexible pavement, repair of deteriorated rigid pavement is a major maintenance task. Major maintenance activities associated with rigid pavement include partial depth pavement repair and full-depth pavement repair.

10.4 Transportation Asset Management

Asset management is the operation, maintenance, upgrade, and disposal of assets in a cost-effective manner. Transportation asset management is the deployment of sound financial and economic principles and processes for effective operation, maintenance, upgrade, and expansion of transportation infrastructure facilities throughout their life cycles.

The strategy and process for asset management often answer questions as listed below.

- What is the current state or condition of an asset?
- What is the desired service and performance condition?
- What are investment approaches available to maintain the desired service and performance condition?
- What is the best investment strategy?

10.4.1 Basic Engineering Economic Principle

The fundamental concept of engineering economics is that money generates money. $1,000 today is not the same as $1,000 one year later. This concept can be illustrated by depositing $1,000 in a savings account now with a 3% annual interest. One year from now, the current $1,000 is worth $1,003.

10.4.1.1 Interest Rate

An interest rate is an amount earned or charged, expressed as a percentage of the principal, for the use of the principle for a period of time. For lenders, the higher the interest rate is, the bigger the return their investments will get. For borrowers, the higher the interest rate is, the more expensive their loans are.

There are two types of interest, simple and compound. In the case of the simple interest, the only earning or cost of using a principle is the original principle. For example, $100 is invested in a 5% simple interest account (e.g., a certificate of deposit account). At the end of the first year, $5.00 interest is earned. For the second year, it is still the original $100, which makes the 5% simple interest.

$$F = P + P \times n \times i_s$$

where
  F is the future value
  P is the present value+
  i_s is the simple interest rate
  n is the length of time

For the above example, the total amount at the end of two years is $F = 100 + 100 \times 5\% \times 2 = 100 + 10 = \$110$.

As opposed to simple interest, compound interest earns interest on both the original principle and the accrued interest. With the above example, if the account is a savings account with a compound 5% interest rate, at the end of the first year, $5 interest is earned based on the $100 principle. At the end of the second year, both the original $100 plus the $5 interest earned in year one will earn interest. So, at the end of the second year, interest earned is $5.25 vs. the otherwise $5.00 in the simple interest case.

$$F = P(1+i)^n$$

where
  F is the future value
  P is the present value
  i is the compound interest rate
  n is the length of time

For the above example, the total amount at the end of the two-year time is $F = 100(1 + 5\%)^2 = \$110.25$

A compound interest rate is often specified on an annual basis, known as the effective annual percentage rate (APR).

If an interest rate is compounded more than once in a year, it can be converted to an effective APR. For example, a 2% monthly compound interest is equivalent to an APR of 26.82%.

$$APR = (1+i)^m - 1$$

TYPES OF INTERESTED RATE

- Simple interest
- Compound interest
- Annual percentage rate (APR)
- Conversion of others to APR

where

APR is the effective annual percentage rate
i=interest rate for one compounding period
m=the number of compounding periods in a year.

With the above example, i=2% with a compounding period of one month, m=12 compounding periods (1 year=12 months), so the APR=$(1+2\%)^{12}-1=26.82\%$.

10.4.1.2 Time Value of Money

Present Value: The concept of present value is to determine the current worth of a future amount of money. It is often used to compare different investment scenarios.

Scenario studies such as determining which investment option is worth the most or least (e.g., $100 today, $110 one year from now, $1,100 ten years from now, and $10 per month for next year) are often carried out with the present value procedure.

Present Worth (P): equivalent amount at present when the time is zero (t=0).
$P/(F, i, n)$ – convert a future value F to a present value P knowing F, i, and n.

$$P = \frac{F}{(1+i)^n}$$

where

F is the future value
P is the present value
i is the compound interest rate
n is the length of time

For example, $1,000 ($F$) promised at the end of three years (n) with an APR of 5% (i) is worth $P = \dfrac{1,000}{(1+5\%)^3} = \$863.84$ now.

$P/(A, i, n)$ – convert equal annual value A to present value P knowing the equal annual A, i, and n.

$$P = A\frac{(1+i)^n - 1}{i(1+i)^n}$$

where

P is the present value
i is the compound interest rate

n is the length of time
A is the annual amount to be saved or borrowed

For example, a payment of $100 every year ($A$) for 15 years ($n$) with an APR of 3% ($i$) is equivalent to $P = \dfrac{100\left[(1+3\%)^{15}-1\right]}{3\%(1+3\%)^{15}} = \$1,194.86$ now.

Future Value: Future value is to convert all past or present values into future values where scenarios can be compared and ranked. The above comparison example covering $100 today, $110 one year from now, $1,100 ten years from now, $10 per month for next year can also be analyzed with the future worth procedure.

Future Worth (F): equivalent future amount at a time of year n ($t=n$).

$F/(P, i, n)$ – convert current value P to future value F knowing P, i, and n.

$$F = P(1+i)^n$$

For example, a deposit of $100 now in an APR of 3% savings account for 15 years is worth $100(1+3\%)^{15} = \$155.80$ in the future.

$F/(A, i, n)$ – convert equal annual value A to future value F knowing A, i, and n.

$$F = A\frac{(1+i)^n - 1}{i}$$

For example, a deposit of $100 every year into an APR of 3% savings account for 15 years is worth $\dfrac{100\left[(1+3\%)^{15}-1\right]}{3\%} = \$1,860$ at the end of the 15th year.

Annual Worth or Cost: The concept of annual worth or cost is to convert present or future worthiness into an annualized term.

$A/(P, i, n)$ – convert present value P to equal annual return A knowing P, i, and n.

$$A = P\frac{i(1+i)^n}{(1+i)^n - 1}$$

For example, with a 3% APR, an investment worth $1,000 now will enable an annual withdraw of $117.18 for each of the next ten years.

$$\frac{1,000\left[3\%(1+3\%)^{10}\right]}{(1+3\%)^{10}-1} = \$117.18$$

Another example is the estimation of monthly payment (equal payment) for a five-year $15,000 loan with an APR of 6%.

First, convert the 6% APR to a compound monthly average interest rate (i) by using the formula $APR = (1+i)^m - 1$. The period for the interest is one month.

There are 12 months (periods) in a year. So, $6\% = (1+i)^{12} - 1$. $i = 0.4866\%$. Total number of monthly payments (n) is 5 years × 12 payment/year = 60 payment.

The equal monthly payment (A) is:

$$A = 15,000 \frac{0.4866\%(1+0.4866\%)^{60}}{(1+0.4866\%)^{60} - 1} = \$288.87/\text{month}$$

$A/(F, i, n)$ – convert future value P to equal annual return A knowing P, i, and n.

$$A = F \frac{i}{(1+i)^n - 1}$$

10.4.2 Equivalence

The ability to compare and contrast the values of different investment alternatives is critical for decision-making. Creating a common benchmark value offers such possibilities. The annual return or annual cost of an alternative is often used as a benchmark value for such comparison.

For alternative analysis about investment and return, three scenarios listed below are often encountered.

a. A single investment right now delivers "n" years of acceptable service.

For example, a full-depth roadway resurfacing project costing $6.2 million now can provide 24 years of acceptable ride quality service.

b. A single investment right now plus a uniform annual investment delivers "k" years of acceptable service.

For example, a full-depth roadway resurfacing project costing $6.1 million now plus an annual maintenance investment of $45,000 can provide 32 years of acceptable ride quality services.

c. A single investment right now delivers "m" years of service + another single investment at the end of "m" years gaining additional "x" years of service.

For example, a full-depth roadway resurfacing project costing $5.1 million now can provide 20 years of acceptable service. At the end of the 20 years, another $2.2 million (present value) of major rehabilitation work can extend the adequate ride quality service for another 14 years.

To determine which alternative offers the best return will require the values of "n," "k," "m," and "x" in addition to an investment interest rate.

The commonly adopted method for such analysis relies on the life cycle "annual cost" concept. With the above example, the life cycle for Alternative a is "n" years. It is "m" years for Alternative b. And it is "$(m+x)$" years for Alternative c (Table 10.2).

Table 10.2 Illustration of Alternative Analysis Based on Annualized Life Cycle Cost

Parameters	Alternative a	Alternative b	Alternative c
Service life (life cycle – years)	24	32	$(20+14)=34$
Interest rate (APR)	3.6%	3.6%	3.6%
Value ($)	$6.2M (present)	$45,000 (present annual (A) maintenance)+$6.1M (present)	$5.1M (present)+$2.2M (present)
Formula	$A = P\,\dfrac{i(1+i)^n}{(1+i)^n - 1}$	$A = A_1 + P\,\dfrac{i(1+i)^n}{(1+i)^n - 1}$	$A = P\,\dfrac{i(1+i)^n}{(1+i)^n - 1}$
Application	$6,200,000 \times \left[\dfrac{3.6\%(1+3.6\%)^{24}}{(1+3.6\%)^{24} - 1}\right]$	$45,000 + 6,100,000 \times \left[\dfrac{3.6\%(1+3.6\%)^{32}}{(1+3.6\%)^{32} - 1}\right]$	$(5,100,000+2,200,000) \times \dfrac{3.6\%(1+3.6\%)^{34}}{(1+3.6\%)^{34} - 1}$
Annual cost	$390,031	$369,114	$427,125

The above analysis shows that investment Alternative 2 is the most economically efficient way and has the lowest cost among all alternatives.

The objective of transportation asset management is to minimize the life cycle cost for managing and maintaining transportation assets such as bridges, tunnels, roads, and other road features. By adopting sound engineering in concert with robust engineering economic analysis, a long-lasting and efficiently invested highway system can be built, operated, and maintained.

10.5 Next Step

Maintenance is not the end of a transportation project development process. Data gathered during maintenance provide needed information to planning for better future condition forecasting and needs assessments. Both the Transportation Improvement Program (TIP) and the long-range plan may be modified based on the latest maintenance data and information.

CONCEPT OF TIME VALUE OF MONEY

- Present value
- Future value
- Equivalence
- Annual cost/return

10.6 Summary

Highway maintenance serves a critical function, ensuring roads are operable under acceptable conditions. While activities such as mowing grass, fixing signs, patching potholes, and plowing snow appear routine, the importance of such actions cannot be overstated. How to maintain and update data and information on what is broken, what needs to be fixed, when to fix, how to prioritize, and how to respond to citizen concerns and complaints are constant challenges faced by highway maintenance departments.

10.7 Discussions

1. Briefly outline the roles of roadside vegetation management.
2. Discuss how you will lead a road maintenance team consisting of mostly high school graduates.
3. Discuss pavement failure types on routes you travel the most.
4. Discuss why state DOT senior leaders are heavily involved in winter snowstorm preparation.

10.8 Exercises

1. Highway maintenance is a minor task for state highway agencies. Yes or No.
2. The goal of highway maintenance is to slow down the deterioration rate of the highway infrastructure. Yes or No.
3. The goal of highway vegetation management is to harvest all the grass for husbandry. Yes or No.
4. Highway right of way vegetation management is an integrated system including vegetation selection, mechanical mowing, animal grazing, and herbicide application. True or False.
5. Highway right of way vegetation prevents soil erosion from both stormwater and wind. True or False.
6. List all roles played by roadside vegetation management.
7. Highway grass and weed mowing is to ensure that vegetation does not grow too high to block a driver's line of sight for safely operating a vehicle. True or False.
8. The most effective way to manage highway vegetation is through herbicide application. True or False.
9. Typically, highway grass mowing does not mow anything lower than 6 inches. True or False.
10. Highway mowing is a dangerous activity. Safety should be emphasized to all involved. True or False.
11. Road salt used for deicing and roadway treatment is generally mined rock salt with sodium chloride (NaCl) as its main component. True or False.
12. Pavement loading distribution mechanisms are the same for flexible and rigid pavements. True or False.
13. Requirements for rigid pavement subgrade preparation are much more "demanding" than that for flexible pavement. True or False.
14. Water seepage under Portland cement pavement is not a big issue, given the load is distributed to a large area. True or False.
15. Freeze and thaw cycles cause a lot of stress to a pavement layer. True or False.
16. IRI is one of many measures used to characterize pavement surface roughness. True or False.
17. Faulting is only related to rigid pavement. True or False.
18. Spalling is only related to rigid pavement. True or False.
19. The freeze-thaw cycle is a significant cause of highway pothole formation. True or False.
20. IRI can be used to characterize pavement surface smoothness of both rigid and flexible pavements. True or False.
21. A salt solution typically has a much lower freezing temperature than pure water. That is why salt is used to deice and prevent a roadway surface from icing up. True or False.

22. 3R projects refer to Resurfacing, Restoration, and Rehabilitation projects. True or False.
23. What is the difference between simple interest and compound interest?
24. A bimonthly (every two months) compound interest rate is 1.2%. What is the equivalent annual percentage interest rate?
25. A pavement management system facilitates the optimum distribution of funds to different roadways to achieve optimum pavement conditions for the entire roadway system. True or False.
26. A $26,000 loan is taken to pave a driveway. The loan has an APR of 3.8% and needs to be paid off in three years with equal monthly payments. What is the monthly payment?
27. What does annualized cost in alternative analysis mean? What kind of decision can you make based on annualized cost information in an alternative study?
28. A newly graduated engineer estimates that he needs to make a monthly payment of "m" $ for "n" years to pay off his $10,000 loan. His friend also has a $10,000 loan with an interest rate doubling of his. The new engineer said that his friend needs to pay twice as much as what he needs to pay every month for "n" years to pay off the loan. Is his statement correct? Explain your answer through a step-by-step analysis.
29. A city engineer needs to make a decision on which alternative she should choose to maintain a major boulevard for its pavement. She can invest $64,000 annually for the next six years to keep the road operable or spend a one-time $280,000 for a major rehabilitation work, which will keep the road operable for eight years. The current market APR is 1.8%. Use the annual cost method to help the engineer to make a rational economic decision.
30. There are all sorts of roadway hazards – fog, flood, objects on the road, disabled vehicles, and animals. List the types of road hazards you have encountered while traveling either as a driver or as a passenger and provide approaches on how to mitigate such hazards if you are responsible for highway maintenance.

Bibliography

Maintaining Flexible Pavements – The Long-Term Pavement Performance Experiment SPS-3 5-year Data Analysis, FHWA, https://www.fhwa.dot.gov/pavement/pub_details.cfm?id=223.

Manual of Practice for an Effective Anti-icing Program: A Guide For Highway Winter Maintenance Personnel, FHWA, https://www.fhwa.dot.gov/reports/mopeap/eapcov.htm.

Materials and Procedures for Repair of Potholes in Asphalt-Surfaced Pavements – Manual of Practice, FHWA Publication Number: FHWA-RD-99-168, FHWA, https://www.fhwa.dot.gov/pavement/pub_details.cfm?id=139.

Index

3C principle 48
4-step travel demand modeling
    mode choice 71
    trip assignment 71, 80
    trip balance 76
    trip distribution 77, 78, 82
    trip generation 71, 72, 82
      - attraction 74
      - production 72
    trip purpose 71, 72, 78
23 United States Code 109 18
85th percentile speed 158, 307

A Policy on Geometric Design of Streets and
    Highways 130, 137, 138
AASHTO 18, 130, 131, 137, 138, 139, 140,
    145, 150, 165, 220, 224, 227, 228
ABET 23
access management 158, 164, 289
Accreditation Board for Engineering and
    Technology 23
aesthetic 125, 126, 137, 149, 174, 314
air quality
    analysis 51
    greenhouse gas 91, 195
    program level 191
    project level 51, 191, 194, 195
    transportation conformity 44, 53, 84, 85,
      88, 90, 92, 98
all red traffic lights 278
American Institute of Certified Planners 24
American National Standard 271
annual average daily traffic 67, 153
annual cost 340
annual percentage rate 337
annual return 339
annual worth or cost 339
asphalt concrete pavement 226

asphalt placement 254
asset management 336
average daily traffic 67
average speed 302

base layer preparation 228, 254
bicycle lane 20, 49, 117, 142, 149, 158, 161, 175
bid
    analysis 265
    awarding 265
    bond 263, 264
    opening 164
    package 262, 263, 264
    package preparation 262
binder course 226
border 143, 144, 231
BPR equation 81
braking capability 131
bridge 221
    abutment 145
    arch 221
    beams 145
    bents 145
    cable-stayed 221, 223
    deck 145, 221
    girder 145, 147, 221, 223
    parapets 147
    piers 145
    scuppers 148
    substructure 145, 212
    superstructure 145, 221
    suspension 221, 223
    truss 221, 223
bulldozer 255
Buy American Act 263

Categorical Exclusions 107
CEI 254

centroid 68, 188
centroid connector 68
certified flaggers 260
CFR 17
class of action determination 106
clear zone 54, 142, 143, 144, 149, 150, 231
CMAQ 4
commitments 20, 106, 107, 164, 172, 220
community cohesion 173
community services 175
compactor 257
competency exam 23
compound interest 337
concentration 3, 191, 192
concrete transport truck 258
condemnation 246, 248
conformity determination 92
conformity lapse 92
Congestion Mitigation and Air Quality 4
constructability review 233
construction
    activity 253, 254
    engineering and inspection 254
    engineers 260
    safety 268
    surveying 236
contamination assessment
    cost reimbursement 196
    incidental releases 196
    liability 196
    liability transferability 196
contamination remediation
    contaminated soil 196, 197
    contaminated water 197
    dewatering 197
continuing resolution 17
contract administration 261
contract changes 266
control surveying 234
cordoned area tolling 299
corrective maintenance 324
crest vertical curve - 3 scenarios 137
cross-section 141
    bridge 147
    rural 143
    urban 143
crosswalk 119
curb 143, 144, 145
curb and gutter 143, 144
curve lenth 123
curve radius 168
cycle length 279

Davis-Bacon commitment 263
DBE 3
dedicated short-range communication 300
defection and cord method 237
deicing 320
design control 150
design exception 231
design speed 158
design traffic 153
design vehicle 150
disadvantaged business enterprise 3
dispersion mechanism
    advection 192
    chemical reaction 193
    diffusion 193
    gravitational settling 193
    radioactive decay 193
    wet and dry deposition 193
dispersion modelling
    Gaussian 194
    worst-case scenario 194
disputes
    arbitration 268
    dispute review board 268
    mediation 268
    mini-trials 268
    negotiation 267
    private judging 268
ditch 116
dredge and fill permit 104
driveway 119
driveway connections 289

EA/FONSI 114
easement 246, 255
EIE 20, 109, 172, 247, 267
EIS
    draft 107
    environmentally preferable alternative 107
    final 107
    preferred alternative 107
    ROD 107
electronic toll collection 299
eminent domain 246, 248, 249
emission inventory 88
empirical design method 227
engineering design surveying 235
environmental impact evaluation 163, 177,
        183, 242
environmental permits 217
ETC 299
eutectic temperature 320

excavator 255
executed contract 265

fair market value 247
farmland 183
FE 207
federal actions 104, 107
federal highway standards 18
federal register 107, 200
Federal Uniform Act 248, 249
federal-aid highway 2
fee-simple purchase 246
final engineering design 207
flagger 259
flexible pavement 226, 325, 330, 336
    alligator cracking 326
    bleeding 329
    block cracking 326
    corrugation deformation 328
    degradation failure 329
    edge cracking 327
    failures 325
    longitudinal cracking 327
    polished aggregate 335
    potholes 330
    raveling 329
    reflective cracking 328
    rutting 328
    shoving 328
    slippage 328
    transverse cracking 328
floodplain 180
fluorescent material 272
flux 192
FMCSA 21
FTA 21
fully-actuated traffic controller 279
future value 337

gravity model 78
guardrail 116

HCM 156
hierarchy of laws and regulations 18
high occupancy
    vehicle 297
    toll lane 298
high-visibility safety apparel 271
highway capacity manual 156
highway maintenance 313
highway noise 183
highway safety improvement program 64

historic and archaeological resources 175
horizontal control survey 234
horizontal curve 124
    curvature 124
    curve central angle 124
    curve length 123
    Long cord 123
    Middle ordinate distance 123
    Point of Curve 123
    Point of Tangent 123
    Point of Tangent Intersection 123
    layout
        broken back 126
        circular 125
        compound 126
        practical design 130, 131
        reverse 127
        spiral 125
HOT 298
HOV 297
HOV lane 297
HSIP 64

ice formation 319
impact
    cumulative 106
    direct 106
    indirect 106
incident management 293
    performance measurement
        incident clearance time 295
        number of secondary crashes 295
        road clearance time 295
initial earth work 254
intelligent transportation system 293
interchange modification report 290
interconnected coordinated traffic controllers 279
interest rate 336
international roughness index 323
international safety equipment association 272
intersection traffic movements 283
interstate maintenance program 4
ITS 293

just compensation 174

labeled stakes 236, 239, 240
land surveying 254
land surveyors 254
land use 49
landscape plan 217
lane delineation mark 117

lane group 284
lead federal agencies 106
legislation
    FAST Act 12
    ISTEA 2
    MAP21 11
    SAFETEA-LU 4
letting date 264
level of service 302
leveling 235, 257
licensing 23
licensing reciprocity 24
lighting plan 208, 214
Load Resistance Factor Design 224
loader 256
local roads 30
LOS 302
lowest responsible bid 265
LRTP 52

major collector roadways 30
Manual on Uniform Traffic Control Devices 214
material failure 329
mechanistic-empirical pavement design 227, 228
median 114, 119, 292
median opening 292
mediation 248, 268
Metropolitan Planning Area 38
Metropolitan Planning Organization 40, 44
Metropolitan Statistical Area 37
Micropolitan Statistical Area 37
mined rock salt 321
minor arterial – other highways 30
minor collector roadways 30
MIRE 54
MiSA 37
Model Inventory of Roadway Elements 54
Model Minimum Uniform Crash Criteria 53
motor grader 257
MOVES 88
MPA 38
MPO
    advisory committee 41
    board 40
    long range plan 44
    products 44
    TIP 44
    voting 41
MSA 37
MUCC 53
MUTCD 214, 228, 229, 230, 231, 232, 233,
    260, 269, 270, 271, 272

NAAQS
    1-hour 191
    CO 191
    maintenance area 40
    non-attainment Area 38
    ozone 39
    PM2.5 39
NAD83 235
National Environmental Policy Act 104
National Highway Freight Program 12
National Highway Performance Program 12
National Highway System 35
National Oceanic and Atmospheric
        Administration 179
National Pollutant Discharge Elimination
        System 181
National Register of Historic Places 176
National Wetland Inventory 177
NAVAD88 235
NCEES 23
NEPA
    Categorical Exclusion 104
    Environmental Assessment 108
    Environmental Impact Statement 106
    Finding of No Significant Impact 104
    Record of Decision 104
NHPP 12
NHS 2
NHTSA 21
noise
    67 dba 190
    antiquity 191
    empirical semi-mechanistic 190
    mechanistic numerical approach 190
    TNM 190
    traffic noise modeling 190
noise abatement criteria
    23 CFR Part 772 183, 184
    activity 184
    activity category 184
    hourly a-weighted 184
noise abatement, insertion loss 191
noise abatement wall 183
    cost 190
    desire 191
    wall height 190
NPDES 217

Occupation Safety and Health
        Administration 269
official appraisal 247
off-tracking 150

origin and destination 68
ozone 38

parks and recreation 176
pavement
    life cycle analysis 227
    marking 214, 228
    smoothness 323
    structural capacity 323
    surface distress 323
pavement management system 324
    reconstruction 324
    rehabilitation 324
paver 258
payment bond 263
pedestrian movements 278
performance bond 263
permit 404 104
permit
    bridge 105
    coast guard 105
    dredge and fill 104
    NPDES 81
permitted movements 280
phase 281, 282
piles
    drilled shafts 145
    precast piles 145
plan and profile 210
planning principle 48
planning regulation 52
Portland cement concrete 258
post-award managing activities 265
post-discharge rate 181
PPM 191, 192, 195
pre-discharge rate 181
preliminary engineering 207
prequalification 263
present value 337
pre-timed coordinated traffic signal
     controller 278
prime coat 329
principal arterial
    interstate 30
    other freeways and expressways 30
    other highways 30
programmatic 108, 109
project cost estimate 93
protected movements 280
PSE 21, 241, 262
public hearings 175
public road length 31

reaction time 130
Real Property Acquisition Policies Act 174,
    246, 247
relocation 248
relocation assistance 248
retroreflective material 272
revenue estimation 51
right of way
    acquisition 246
    map 246
    plan 247
    surveying 235
rigid pavement 331
    corner cracks 333
    D cracking 332
    faulting 334
    linear cracking 331
    polished aggregate 335
    pumping 334
    scaling 333
    shrinkage cracking 332
    spalling 334
    warping cracks 332
road salt 321
road weather 321
roadside vegetation 313
    herbicide application 314
    mowing 315
    mowing safety 315
    nonselective herbicide 318
    normal mowing 316
    selective herbicide 318
    special mowing 316
    weeds control 314, 316
    wildfire control 314
roadway classification 29
    administrative approach 35
    functionality based 30
    ownership based 29
roadway plan 208
roadway signs 118
    overhead 117
    roadside 118
    variable 118
roller 257
ROW 246
rumble strip 115
rural area 37

SAFETEA-LU Title 5 4
safety data analysis 53
sag vertical curve - 3 scenarios 130

Section 4(f) 177
Section 404 permit 179
semi-actuated traffic control 279
shoulder 114
SHPO 176
side walk 117
sight distance
    downhill 132
    eyesight height 141
    measuring 141
    object height 141
    stopping 129
    uphill 132
sign and pavement marking plan 214
signal phasing 281
    2-phase control 281
    3-phase control 282
    4-phase control 283
    split 2-phase control 282
simple interest 336
SIP 38, 40, 51
skid resistance 224
slope stake 236, 240
snow and ice control 320
snow removal 320
sound
    decibels 187
    energy 187
    frequency 186
    L10 188
    Leq 188
    loudness 187
    octave band 186
    pressure 187
    loudness
        A-weighting 187
        C-weighting 187
        Z weighting 188
    propagation
        atmospheric absorption 188
        barrier attenuation 189
        geometric attenuation 188
        terrain ground surface bouncing 189
    speed 186
    wavelength 186
spot speed 302
staking process 236
standard construction material 219
standard pay items 219
State Historic Preservation Officer 176
state implementation plan 38, 85
statewide transportation improvement
        programs 43

station 121
station equation 121
STIP 43
Stochastic User Equilibrium Method 81
stormwater
    detention 120
    flooding 120
    quality 180, 181
    retention 120, 180, 181
storm-water runoff 182
Strategic Defense Highway Network 35
Strategic Highway Safety Plan (SHSP) 53
superelevation 134
    friction coefficient 134
    maximum 140
    normal crown 135, 136
    practical design 139
    radius 135
    runoff 136
    runout 136
    transition rate 136
    transverse friction force 133
surface friction property 323
surface smoothness 323
survey stakes 236
system optimum solution 81

tack coat 226
tangents 236
temporary traffic control zone
    activity work area 271
    advance warning area 270
    termination area 271
    transition area 271
Titles 23 and 49 3
TMA 38
traffic
    D_{30} 155
    design directional factor 155
    design hour factor 155
    directional design hour volume 155
    generalized level of service 157
    K_{30} 155
    maximum practical service flow rate 155
    maximum service flow rate 156
    number of lanes 155
traffic analysis zones 68
    Macro TAZs 68
    Meso TAZs 68
    Micro TAZs 68
traffic light 278
Traffic Records Coordinating Committee 64
traffic signal control system 278

traffic signal timing 286
  critical lane group 284
  dilemma zone 284
  effective green 284
  minimum green time 285
trained flagger 260
transportation conformity 84
Transportation Curriculum Coordination
    Council 259
transportation demand management 297
Transportation Improvement Programs 44
Transportation Management Area 38
travel cost factor 78
travel demand analysis 64
  corridor 65
  links 66, 67
  network 66
  nodes 66, 67
travel time index 302
traveled way 114
TRCC 64
triangulated irregular network 235
Truck National Network 35
turning vehicle 133
turning vehicle, force 133
two-way left-turn lane 292
typical section 148

Uniform Relocation Assistance 246
University Transportation Centers 17
urban area 37
urban cluster 37
urbanized area 37
U.S. Army Corps of Engineers 104
U.S. Coast Guard 105
U.S. DOT organizational structure 44
U.S. Fish and Wildlife Service 179

USC 18
user equilibrium method 81
UTC 17
utility
  accommodate 197
  adjustment 197
  local government 198
  reimbursement agreement 198
  removal 198
utility assessment 197

variable tolling lanes 198
vehicle classification 300
vertical control survey 235
vertical curve 127
  curve length 128
  elevation 128
  vertical curve formula 128
  Vertical Point of Curve 127
  Vertical Point of Intersection 127
  Vertical Point of Tangent 127
VOC 85, 88

warranty bond 263
wearing course 226
wetland
  boundary 178
  impact 178, 179
  mitigation 179
  mitigation bank 179
  mitigation ratio 179
  quality 177
wildlife and habitat
  action area 179
  incidental take 180
work zone safety 269
workforce 18, 259, 263

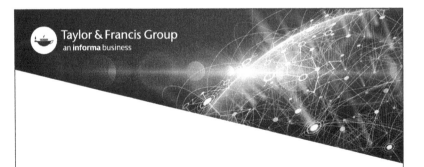

Printed in the United States
By Bookmasters